THE ART OF GLASS-BLOWING

LECTOR HOUSE PUBLIC DOMAIN WORKS

THE ART OF GLASS-BLOWING

T. P. DANGER

ISBN: 978-93-90015-09-2

Published: 1831

LECTOR HOUSE LLP
E-MAIL: lectorpublishing@gmail.com

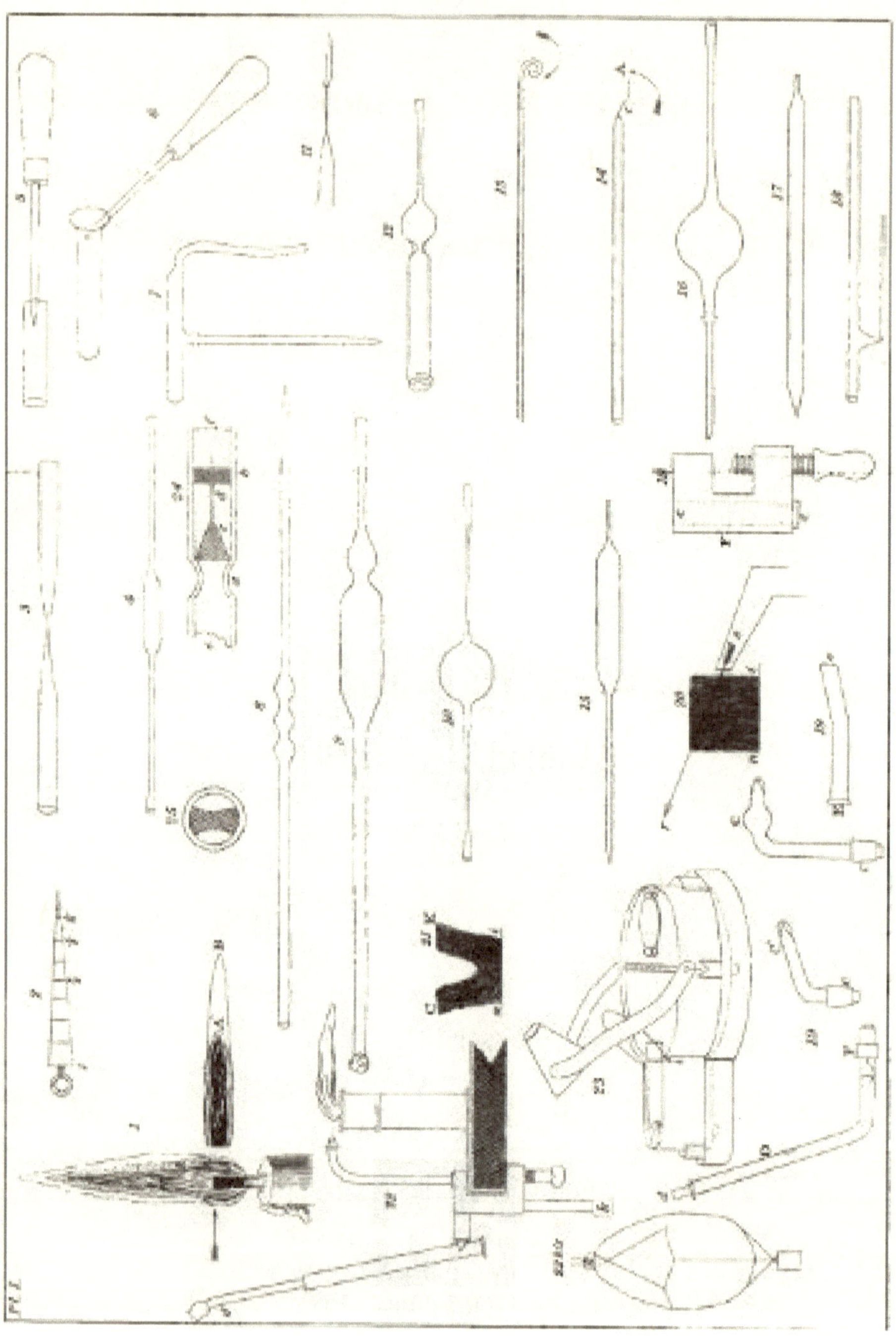

THE ART OF GLASS-BLOWING,

OR PLAIN INSTRUCTIONS FOR MAKING THE CHEMICAL AND PHILOSOPHICAL INSTRUMENTS WHICH ARE FORMED OF GLASS;

SUCH AS

BAROMETERS, THERMOMETERS, HYDROMETERS,

HOUR-GLASSES, FUNNELS, SYPHONS,

TUBE VESSELS FOR CHEMICAL EXPERIMENTS, TOYS FOR RECREATIVE PHILOSOPHY, &C.

BY

T. P. DANGER

A FRENCH ARTIST.

ILLUSTRATED BY ENGRAVINGS.

1831.

PL. 2.

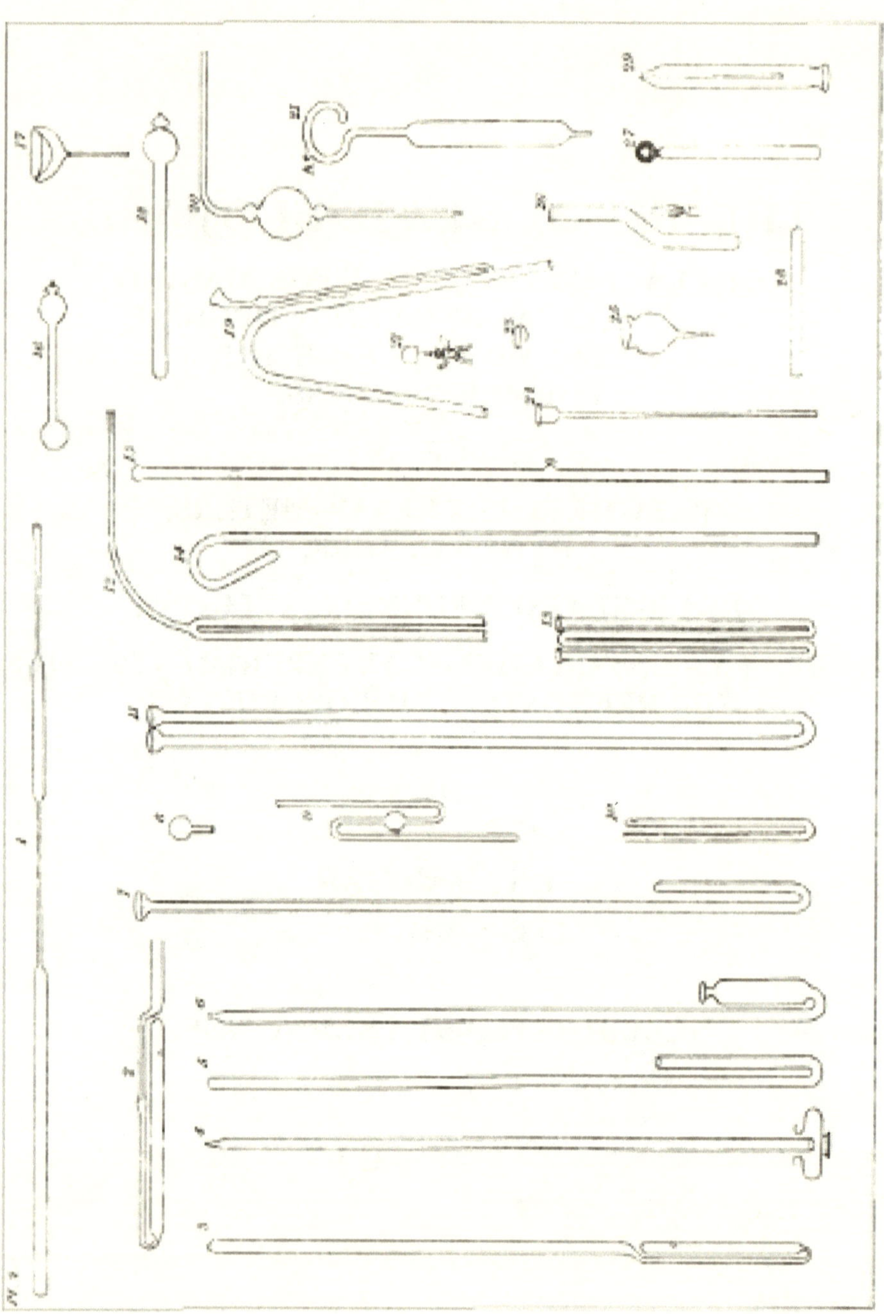

PL. 3.

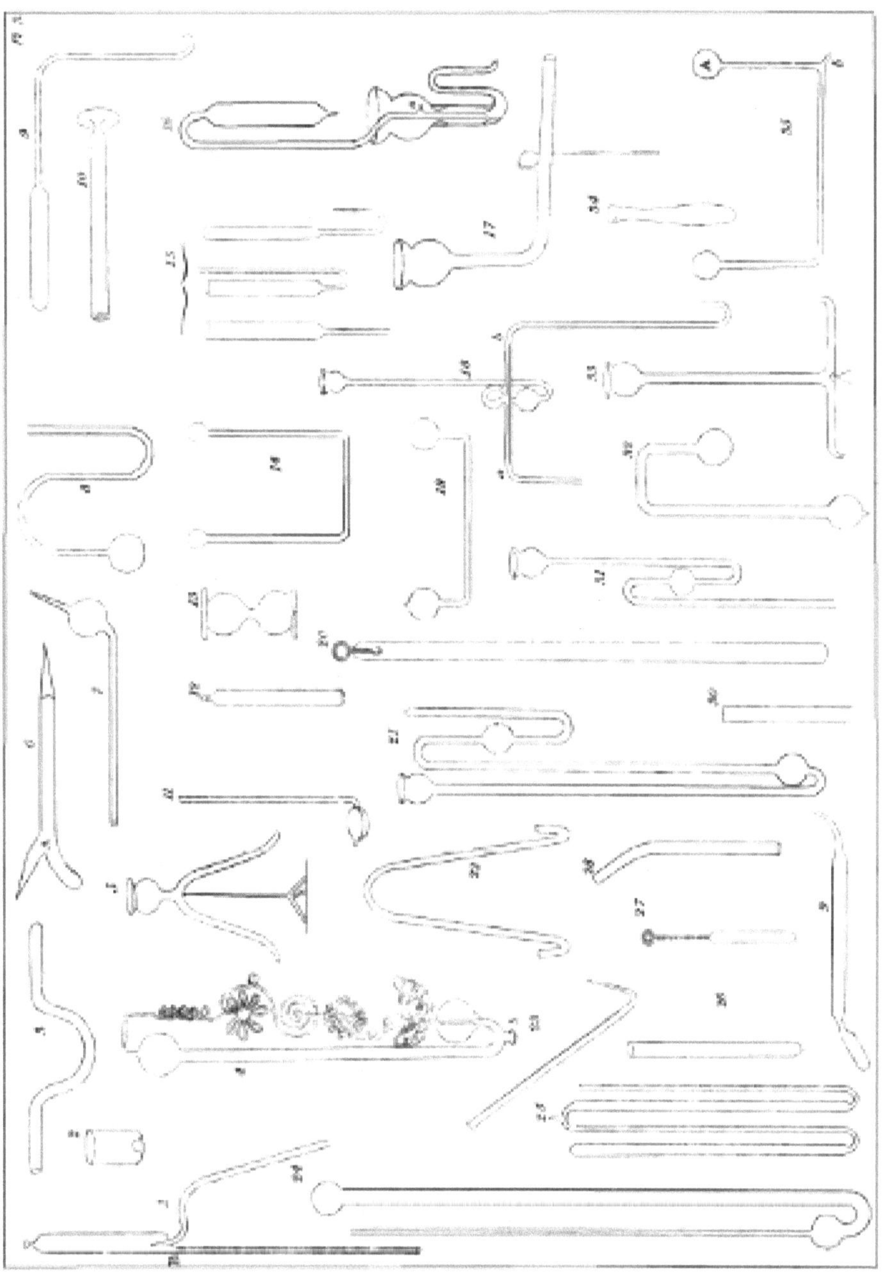

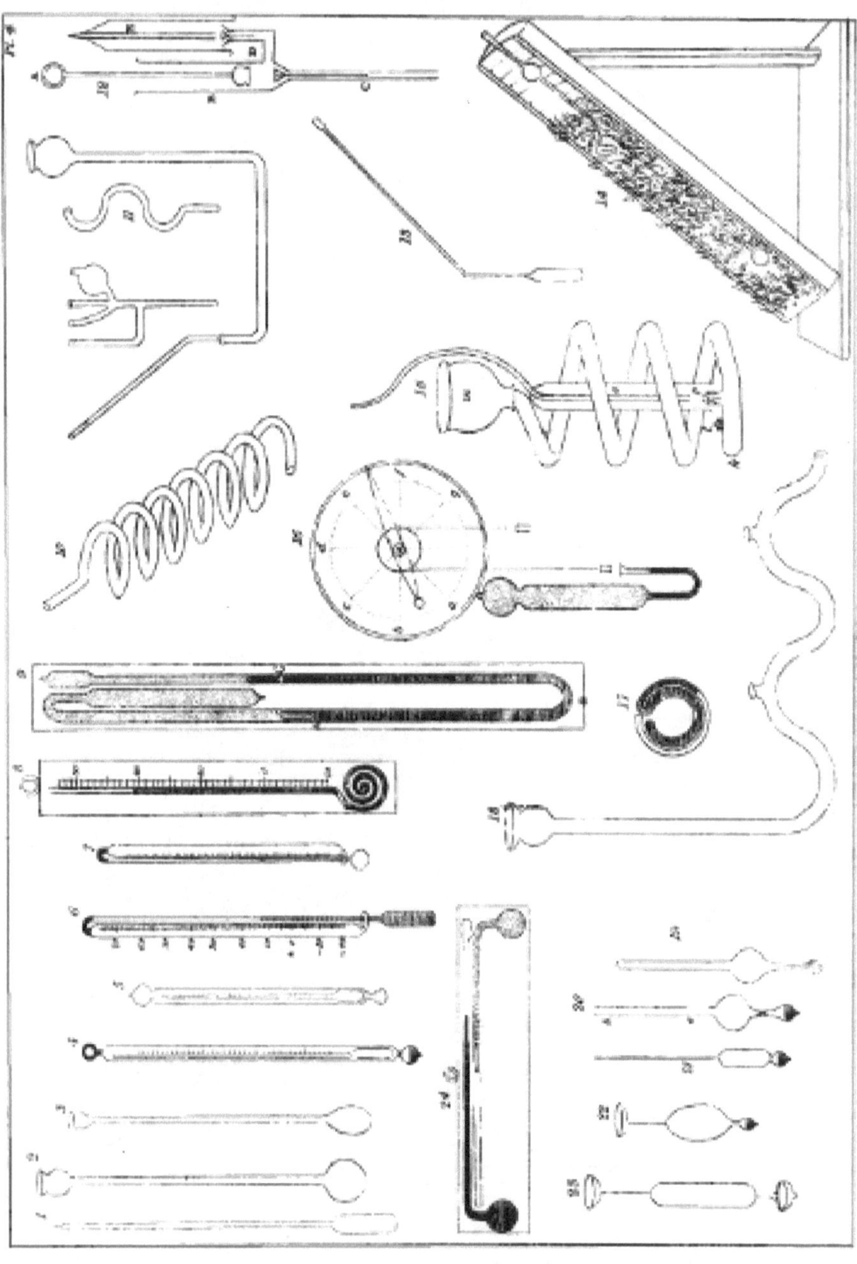

TRANSLATOR'S PREFACE.

The scientific instruments prepared by the glass-blower are numerous and highly useful: barometers, thermometers, syphons, and many other vessels constructed of tubes, are indispensable to the student of physics or chemistry. Some of these instruments are high in price, and liable to frequent destruction; and those by whom they are much employed are subject to considerable expense in procuring or replacing them. It is therefore advisable that he who desires to occupy himself in the pursuit of experimental science, should know how to prepare such instruments himself; that, in short, he should become his own glass-blower. "The attainment of a ready practice in the blowing and bending of glass," says Mr. Faraday, "is one of those experimental acquirements which render the chemist most independent of large towns and of instrument-makers."

Unquestionably the best method of learning to work glass is to obtain personal instructions from one who is conversant with the art: but such instructions are not easily obtained. The best operators are not always the best teachers; and to find a person equally qualified and willing to teach the art, is a matter of considerable difficulty. In large towns, workmen are too much engaged with their ordinary business to step aside for such a purpose; and in small towns glass-blowers are seldom to be found. In most cases, also, they are too jealous of their supposed *secrets* to be willing to communicate their methods of operating to strangers, even when paid to do so.

The following Treatise is a free translation of *L'Art du Souffleur à la Lampe, par* T. P. DANGER. The author is employed, in Paris, in preparing glass instruments for sale, and in teaching others the art of preparing them. He has presented in this work the most minute instructions for the working of glass which have ever been offered to the public. The general processes of the art are so fully explained, and the experimental illustrations are so numerous, that nothing remains except the reducing of these instructions to practice to enable the student to become an adept in the blowing of glass. I trust that, in publishing this work in an English dress, I may be considered as aiding in some degree the progress of physical science.

This work contains a description of a cheap blowpipe and a very convenient lamp; both of them the invention of the author: but any other kind of lamp or blowpipe may be employed instead of these. The reader who wishes for a description of the blowpipes generally employed in England, may consult Mr. GRIFFIN's *Practical Treatise on the Use of the Blowpipe in Chemical and Mineral Analysis.*

London, September 1831.

AUTHOR'S PREFACE.

The flame of a lamp, or candle, condensed and directed by a current of air, is exceedingly useful in a great number of arts. The instrument which is employed to modify flame is the Blowpipe. This is an indispensable agent for jewellers, watch-makers, enamellers, glass-blowers, natural philosophers, chemists, mineralogists, and, indeed, for all persons who are occupied with the sciences, or their application to the arts. Its employment offers immense advantages in a multitude of circumstances; and the best method of making use of so powerful an agent ought to be well known to every person who is likely to be called upon to adopt it.

Students, especially those who desire to exercise themselves in chemical manipulation, must feel the want of a simple and economical process, by means of which they could give to glass tubes, of which they make great use, the various forms that are necessary for particular operations. How much reason have they to complain of the high price of the instruments of which they make continual use! The studies of a great number are shackled from want of opportunity to exercise themselves in manipulation; and many, not daring to be at the expense of a machine of which they doubt their ability to make an advantageous use, figure to themselves the employment of the glass-blower's apparatus as being beset with difficulties, and so rest without having even an idea of the numberless instruments which can be made by its means.

Many persons would very willingly occupy their leisure time in practising the charming art of working glass and enamels with the blowpipe; but the anticipated expense of the apparatus, and the difficulties which they imagine to foresee in the execution of work of this kind, always repels them.

The new species of blowpipe which we have offered to the public, and which has received the approbation of the Society for the Encouragement of Arts, obviates all these inconveniences: its moderate price, its portability, and the facility with which it can be used, adapt it to general employment.

But we should not believe that we had attained the end which we had proposed to ourselves if we had not placed young students in a situation to repeat at their own houses, at little cost, and with the greatest facility, the experiments which are necessary to familiarise them with the sciences. It is with such a view that we present to them this little Treatise, which is destined to teach them the simplest, the most expeditious, the least expensive, and the most effectual methods of constructing themselves the various instruments which they require in the prosecution of their studies.

The word *glass-blower*, generally speaking, signifies a workman who occupies himself in making of glass and enamel, the instruments, vessels, and ornaments, which are fabricated on a larger scale in the glass-houses: but the domain of the sciences having laid the art of glass-blowing under contribution, the artists of the lamp have divided the labours thereof. Some apply themselves particularly to the construction of philosophical and chemical instruments; others occupy themselves with little ornamental objects, such as flowers, &c.; and, among the latter, some manufacture nothing but pearls, and others only artificial eyes. Finally, a few artists confine themselves to drawing and painting on enamel, which substance is previously applied to metallic surfaces by means of the fire of a muffle.

As we intend to treat separately of these different branches of the art, we commence with that of which the manipulation is the simplest.

Paris, 1829.

Contents.

PART I.—INSTRUCTIONS FOR THE DETECTION OF MINERAL POISONS IN VEGETABLE OR ANIMAL MIXTURES.—Copper, Lead, Antimony, Arsenic, Mercury, Iron, Barytes, Lime, Alumina, Potash, Soda, Sulphuric Acid, Nitric Acid, Muriatic Acid.

PART II.—INSTRUCTIONS FOR THE EXAMINATION OF ARTICLES SUPPOSED TO BE ADULTERATED.—Alcohol, Ale, Anchovy Sauce, Arrow-Root, Beer, Brandy, Bread, Calomel, Carmine, Cayenne Pepper, Cheese, Chocolate, Chrome Yellow, Cinnamon, Cloves, Cochineal, Coffee, Confectionery, Crabs' Eyes, Cream, Cream of Tartar, Epsom Salts, Flour, Gin, Gum Arabic, Spirits of Hartshorn, Honey, Hops, Ipecacuanha, Isinglass, Ketchup, Lakes, Leeches, Lemon Acid, Litharge, Magnesia, Milk, Mushrooms, Mustard, Olive Oil, Parsley, Pepper, Peruvian Bark, Pickles, Porter, Red Oxide of Mercury, Rhubarb, Sal Ammoniac, Salt, Saltpetre, Soap, Soluble Tartar, Spanish Liquorice, Spirits, Sugar, Sulphur, Tamarinds, Tapioca, Tartaric Acid, Tartar Emetic, Tea, Ultramarine, Verdigris, Vermilion, Vinegar, Volatile Oils, Wax, White Lead, Wine, Water,(including directions for testing the purity of all descriptions of Rain, River, or Spring Water.)

PART III.—INSTRUCTIONS FOR THE PREPARATION OF THE TESTS EMPLOYED IN DOMESTIC CHEMISTRY AND FOR THE PERFORMANCE OF VARIOUS CHEMICAL OPERATIONS; WITH DESCRIPTION OF THE GLASSES AND APPARATUS PROPER TO BE EMPLOYED.

** The work is written in a popular manner, and intended for the use of Families, Publicans, Wine and Spirit Merchants, Oilmen, Manufacturers, Apothecaries, Physicians, Coroners, and Jurymen.—Price Three Shillings.

CONTENTS

Page

I. Instruments Employed in Glass-Blowing. 1

   • The Blowpipe.. 1

   • The Glass-Blower's Table. 2

   • The Eolipyle.. 3

   • Blowpipe With Continued Current. 3

   • The Lamp. 4

   • The Candlestick. 4

   • Combustibles. 5

      » Oil, Tallow, &c. 5

      » The Wicks. 5

      » Relation between the diameters of the beaks of the blow-pipe, and the wicks of the lamp. 6

II. Preliminary Notions of the Art. 8

   • The Flame. 8

   • Places Fit To Work In. 9

   • Means Of Obtaining A Good Fire. 9

   • Choice And Preservation Of Glass. 10

   • Preparation Of Tubes Before Heating Them. 12

   • Method Of Presenting Tubes To The Fire, And Of Working Them Therein.. 12

III. Fundamental Operations in Glass-Blowing. 15

   • 1.—Cutting. 15

   • 2.—Bordering. 17

   • 3.—Widening. 17

- 4.—Drawing Out. 18
- 5.—Choking.. 18
- 6.—Sealing. 18
- 7.—Blowing.. 19
- 8.—Piercing. 22
- 9.—Bending.. 23
- 10.—Soldering. 24

IV. Construction of Chemical and Philosophical Instruments. 26
- Adapters. 26
- Cryophorus 26
- Archimedes's Screw. 27
- Barker's Mill. 27
- Barometers. 28
    - »   Cistern Barometer.. 28
    - »   Dial (or Wheel) Barometer. 28
    - »   Syphon Barometer.. 28
    - »   Stop-Cock Barometer. .28
    - »   Compound Barometers.. 28
    - »   Gay Lussac's Barometer. 29
    - »   Bunten's Barometer.. 29
    - »   Barometer pierced laterally for Demonstrations.. 29
- Bell-Glasses For Experiments.. 29
- Blowpipe. 30
- Capsules.. 30
- Cartesian Devils. 31
- Communicating Vases.. 31
- Dropping Tubes. 31
- Fountains. 32
    - »   Fountain of Circulation.. 32
    - »   Fountain of Compression.. 32
    - »   Intermitting Fountain.. .32
    - »   Hero's Fountain.. 33
- Funnels. 33
- Hour-Glasses. 33

- Hydraulic Ram. 34
- Hydrometers. 34
  » Baumé's Hydrometer. 34
  » Nicholson's Hydrometer. 35
  » Hydrometer with two Branches. 35
  » Hydrometer with three Branches. 35
  » Hydrometer with four Branches. 35
- Manometers. 36
- Mariotte's Tube. 36
- Phosphoric Fire-Bottle. 36
- Pulsometer. 36
- Pump. 36
- Retort For Chemical Experiments. 37
- Tubulated Retort. 37
- Thermoscope. 37
- Rumford's Thermoscope. 37
- Syphons. 38
- Spoons. 38
- Spirit Level. 38
- Test Glass With A Foot. 39
- Thermometers. 39
  » Ordinary Thermometer. 39
  » Dial Thermometer. 40
  » Chemical Thermometer. 40
  » Spiral Thermometer. 41
  » Maximum Thermometer. 41
  » Minimum Thermometer. 41
  » Bellani's Maximum Thermometer. 42
  » Differential Thermometer. 42
- Tubes Bent For Various Purposes 42
- Vial Of The Four Elements. 43
- Water Hammer. 43
- Welter's Safety Tubes. 44

V. Graduation of Chemical and Philosophical Instruments. 45

- Of The Substances Employed In The Preparation Of These Instruments. 45
- Of Graduation In General. 46
- Examination Of The Bore Of Tubes. 46
- Division Of Capillary Tubes Into Parts Of Equal Capacity. 46
- Graduation Of Gas Jars, Test Tubes, &C.. 47
- Graduation Of Hydrometers. 48
- Graduation Of Barometers. 49
- Graduation Of The Manometer.. 51
- Graduation Of Thermometers. 51
- Graduation Of Rumford's Thermoscope. 54
- Graduation Of Mariotte's Tube.. 54

THE ART OF GLASS-BLOWING.

I.

INSTRUMENTS EMPLOYED IN GLASS-BLOWING.

On seeing, for the first time, a glass-blower at work, we are astonished at the multitude and the variety of the modifications to which he can make the glass submit. The small number and the simplicity of the instruments he employs, is also surprising. The blowpipe, or, in its place, the glass-blower's bellows and a lamp, are indeed all that are indispensable.

THE BLOWPIPE.

Originally, the blowpipe was only a simple, conical tube, more or less curved towards its point, and terminated by a very small circular opening. By means of this, a current of air was carried against the flame of a candle, and the inflamed matter was directed upon small objects, of which it was desirable to elevate the temperature. Workers in metal still derive immense advantages from the use of this little instrument: they employ it in the soldering of very small articles, as well as for heating the extremities of delicate tools, in order to temper them. But since the blowpipe has passed into the hands of mineralogical chemists, its form has been subjected to a series of very curious and important modifications. In spite, however, of these ameliorations, which rendered the instrument better adapted for the uses to which it was successively applied, we are far from having drawn from it all the advantages to which we might attain, were its employment not as fatiguing as it is difficult. We require no other proof of this than the small number of those who know well how to make use of the blowpipe.

The most economical blowpipe is a tube of glass, bent near one end, and pointed at its extremity. A bulb is blown near that part of the tube which corresponds with the curvature (pl. 3, fig. 7.) This bulb serves as a reservoir for moisture deposited by the air blown into the tube from the mouth. If you employ a tube without a bulb, the moisture is projected in drops into the flame, and upon the objects heated by it—an effect which is very inconvenient in practice. To put this instrument into action, accustom yourself to hold the mouth full of air, and to keep the cheeks well inflated, during a pretty long series of alternate inspirations and expirations; then, seizing lightly with the lips the mouth of the blowpipe, suffer the air compressed by the muscles of the cheeks, which act the part of a bellows, to escape by the beak of the blowpipe, which you will be able to do without being put to the least inconvenience with regard to respiration. When the air contained in the mouth is pretty nearly expended, you must take advantage of an inspiration, to inflate the lungs afresh; and thus the operation is continued. You must never blow through the tube by means of the lungs; first, because air which has been in the lungs is less proper for combustion than that which has merely passed through the nose and mouth; secondly, because the effort which it would be necessary to make, to sustain the blast for only a short time, would by its frequent repetition become very injurious to your health.

The jet of flame produced by the mouth-blowpipe can only be used to heat small objects: when instruments of a considerable bulk have to be worked, it is customary to employ the *lamp*, or *glass-blower's table*.

THE GLASS-BLOWER'S TABLE.

Artists give this name to an apparatus which consists of the following articles:—

1. A *Table*, below which is disposed a *double bellows*, capable of being put in motion by means of a pedal. This bellows furnishes a continued current of air, which can be directed at pleasure by making it pass through a tube terminating above the table in a sharp beak. The bellows with which the glass-blower's tables are commonly furnished have very great defects. The irregular form which is given to the pannels diminishes the capacity of the instruments, without augmenting their advantages. If we reflect an instant on the angle, more or less open, which these pannels form when in motion, we instantly perceive that the weight with which the upper surface of a bellows is charged, and which always affords a vertical pressure, acts very unequally on the arm of a lever which is continually changing its position. This faulty disposition of the parts of the machine has the effect of varying every instant the intensity of the current of air directed upon the flame. All these inconveniences would disappear, were the upper pannel, like that in the middle, disposed in such a manner as to be always horizontal. It ought to be elevated and depressed, in its whole extent, in the same manner; so that, when charged with a weight, the pressure should be constantly the same, and the current of air uniform.

2. A *lamp*, of copper or tin plate.—The construction of this article, sufficiently imperfect until the present time, has varied according to the taste of those who have made use of it. We shall give, farther on, the description of a lamp altogether novel in its construction.

3. The glass-blower's table is generally furnished with little *drawers* for holding the tools employed in modelling the softened glass. Careful artists have the surface of their table coated with sheet iron, in order that it may not be burned by the hot substances that fall, or are laid upon it. As glass-blowers have frequent occasion to take measures, it is convenient to have the front edge of the table divided into a certain number of equal parts, marked with copper nails. This enables the workman to take, at a glance of the eye, the half, third, or fourth of a tube, or to give the same length to articles of the same kind, without having perpetual re-

course to the rule and compasses. But when it is desirable to have the tubes, or the work, measured with *greater exactness* than it can be measured by this method, the rule and the compasses can be applied to.

THE EOLIPYLE.

We shall merely make mention of this instrument. It is a globular vessel, commonly formed of brass. If filled with a very combustible liquor, such as alcohol, and strongly heated, it affords a rapid current of vapour, which, if directed by means of a fine beak into the middle of a flame, produces the same effect as the air which issues from a blowpipe. The eolipyle is a pretty toy, but not a good instrument for a workman, its action being too irregular.

BLOWPIPE WITH CONTINUED CURRENT.

It is after having, during a long period, made use of the instruments of which we have spoken, and fully experienced their inconveniences, that, aware of the indispensable necessity for such instruments in the arts and sciences, we have thought it our duty to make known to the public *a New Apparatus*, which is, not only calculated to fulfil the same purposes, but presents advantages which it is easy to appreciate. The price of it is only the sixth part of that of the glass-blower's table[1]. It is very portable, and capable of being attached to any table whatever. It unites the advantages of not fatiguing the workman, of leaving his hands free, and of rendering him absolute master of the current of air, which he can direct on the flame either of the lamp or the candle,—advantages which are not offered in the same degree even by the table of the glass-blower.

The instrument which we have presented is, properly speaking, nothing but a simple blowpipe, C, (pl. 1, fig. 19) communicating with a bladder, or leather bag, fixed on E, which is kept full of air by means of a bent tube, D, through which the operator blows occasionally with the mouth. This tube is closed at its inferior extremity, F, by a valve, which permits the passage of air into the reservoir, but not of its return, so that the air can only escape by the beak of the blowpipe.

The valve at F is constructed in the following manner:—At about two inches from the end of the tube a contraction is made, as represented at *a*, pl. 1, fig. 24. This reduces the internal diameter of the tube about one-third. A small conical piece of cork or wood is now introduced into the tube in the manner represented by *c*. The base of the cone must be large enough to close the tube at the point where it is contracted; it must, however, not be so large as to close the tube at the wide part. A brass pin is inserted in the point of the cone, as is shewn in the figure. Between the cone and the end of the tube, the piece of wood, *b*, is fixed; the shape of this piece of wood is best shewn by figure 25, on the same plate. There is a hole in the centre, in which the pin of the cork cone can move easily. The cone or valve is therefore at liberty to move between the contraction *a*, and the fixture *b*. Consequently, when air is blown into the tube at *e*, the valve is forced from the con-

[1] In Paris, the blowpipe which is here described is sold for six francs (five shillings English); or, with the improved lamp and candlestick, twelve francs.

traction, falls into the position indicated by the dotted lines *d*, and allows the air to pass by its sides. When, on the contrary, the operator ceases to blow, the valve is acted upon by the air in the bladder, which, pressing back at *f*, drives the valve close against the contraction, and effectually closes the aperture. A slight hissing is heard, but when the contraction is well made, and the cork is good, an extremely small quantity of air escapes.

The workman, seated before the table where he has fixed his instrument, blows from time to time, to feed the reservoir or bladder, which, being pressed by a system of strings stretched by a weight, produces an uniform current of air. The force of this current of air can be modified at pleasure, by pressing the reservoir more or less between the knees. (Fig. 22 represents a blowpipe complete, formed not of glass, but of brass tubes. Fig. 22, *bis*, represents the bladder or reservoir appertaining to this blowpipe.)

M. Gaultier de Claubry, who was charged by the Committee of Chemical Arts of the Society of Encouragement (of Paris) to make a report on this instrument, was astonished at the facility with which the author, in his presence, reduced the oxide of cobalt to the metallic state, and fused the metal to a globule; an experiment which even M. Berzelius could not perform with the simple blowpipe, since he expressly says, in his work on that instrument, that oxide of cobalt suffers no change when heated before the blowpipe. The results obtained with cast iron, oxide of tin, &c.—experiments which are exhibited every day at the public lectures given by the author—evidently prove the superiority of this apparatus over all the blowpipes that have hitherto been contrived.

A detailed account of the glass tubes belonging to this improved blowpipe will be found in the fourth part of this work, at the article *Blowpipe*.

THE LAMP.

While occupied in rendering popular, if we may so speak, the use of the *blow-pipe*—an instrument which is so advantageous in a great number of circumstances—we have also endeavoured to improve *the lamp*, which has, until the present time, been used by all those who employ the glass-blower's table. The lamp which we recommend (pl. 1, fig. 23) is of a very simple construction. It possesses the advantages of giving much less smoke than the old lamp, and of being cleaned with the greatest facility. It also gives sensibly more heat; because the portion of flame which, in the common lamps, rises perpendicularly, and is not used, is, in this case, beaten down by a cap or hood, and made to contribute to the force of the jet. This cap also keeps the flame from injuring the eyes of the operator, and destroys the smoke to such an extent, that the large hoods with which glass-blowers commonly garnish their work table, to carry off the smoke, become unnecessary. This is a peculiar advantage in the chamber of a student, where a large hood or chimney can seldom be conveniently prepared.

THE CANDLESTICK.

For mineralogical researches, chemical assays, and the soldering of small objects, as in jewellery, we recommend the use of a little candlestick, which, by means of a spring fixed to the bottom, maintains the candle always at the same height. A reservoir, or shallow cup, formed at the top of the candlestick, to hinder the running away of the tallow or wax, allows the operator to consume the fragments of tallow or grease which are ordinarily lost in domestic economy. There is a little hole in the centre of the cup or upper part of the candlestick, through which the wick of the candle passes. *o*, pl. 1, fig. 22, is a representation of this candlestick.

COMBUSTIBLES.

Oil, Tallow, &c.

Among the substances which have been employed to feed the fire of the glass-blower's lamp, those to which the preference is to be given are wax, olive oil, rape oil, poppy oil, and tallow. Animal oils, such as bone oil and fish oil, are much esteemed by some glass-blowers, who pretend that with these substances they obtain better results than with other combustibles. Nevertheless, animal oils, generally speaking, do not give so much heat as purified rape oil, while they exhale an odour which is extremely disagreeable.

As to alcohol, which is sometimes used with the eolipyle, its combustion furnishes so feeble a degree of heat that its employment cannot be recommended.

Purified rape oil is that of which the use is the most general. Next to olive oil and wax, it affords the greatest heat, and the least smoke. But, in a word, as in the working of glass, the operator has more need of a bright flame without smoke, than of a high temperature, any combustible may be employed which is capable of furnishing a flame possessing these two qualities. The vegetable oils thicken, and suffer alterations more or less sensible, when they are long exposed to the action of the air. They should be chosen very limpid, and they may be preserved in that state by being enclosed in bottles, which should be kept quite full and well corked.

The Wicks.

There has never been any substance so generally used for wicks as cotton; some glass-blowers, indeed, have employed wicks of asbestus, but without deriving from them the advantages which might have been expected; the greater number, therefore, keep to cotton.

But it has been observed that cotton which has been for some time exposed to the air no longer possesses the good properties for which glass-blowers esteem it. The alteration of the cotton is probably brought about by the dust and water which the air always holds in suspension. Such cotton burns badly, forms a bulky coal, and permits with much difficulty the capillary ascension of the liquid which serves to support the flame; so that it is impossible to obtain a good fire, and necessary to be incessantly occupied in snuffing the wick. Cotton is equally subject to alteration when lying in the lamp, even though impregnated with oil. You should avoid making use of wicks that are too old. When you foresee that you will remain a long time without having occasion to employ the lamp, pour the oil into a bottle, which can be corked up, and let the wick be destroyed, previously squeezing from it the oil which it contains.

It is indispensable to make use of none but new and good cotton; it should be clean, soft, fine, and not twisted. It is best to preserve it in boxes, after having folded it in many double papers, to exclude dust and moisture. When you wish to make wicks, take a skein of cotton and cut it into four or six pieces, dispose them side by side in such a manner as to make a bundle, more or less thick, and eight or ten inches in length; pass a large comb lightly through the bundle, to lay the threads even, and tie it gently at each end, to keep the threads from getting entangled.

Relation between the diameters of the beaks of the blowpipe, and the wicks of the lamp.

We believe that we cannot place better than here a few observations respecting the size of the opening in the beak of the blowpipe, considered in relation to the size of the wick of the lamp. These observations will probably be superfluous to those who are already conversant with the use of the blowpipe; but as every thing is interesting to beginners, who are frequently stopped in their progress by very slight difficulties, and as this Treatise is particularly designed for beginners, we do not hesitate to enter into the minutest details on subjects which we deem interesting.

The point of your blowpipe should be formed in such a manner, that you can fix upon it various little beaks or caps, the orifices in which, always perfectly round, ought to vary in size according to the bulk of the flame upon which you desire to act. You cannot, without this precaution, obtain the maximum of heat which the combustion of the oil is capable of affording. This employment of little moveable caps offers the facility of establishing a current of air, greater or smaller, according to the object you wish to effect; above all, it allows you to clean with ease the cavity or orifice of the beak, as often as it may be necessary.

These caps can be made of different materials. It is most advisable to have them made of copper or brass; those which are formed of tin plate (white iron), and which are commonly used in chemical laboratories, are the worst kind of all. They soon become covered with grease or soot, which either completely closes up the orifices, or, at least, very soon alters the circular form which is necessary to the production of a good fire. Glass caps are less liable to get dirty, and are much cheaper than the above; but, on the other hand, they have the disadvantage of being easily melted. This can to a certain extent be remedied by making the points of very thick glass, and by always keeping them at some distance from the flame. Moreover, as you can make them yourself when you are at leisure, their use is very commodious. If they are to be used with the blowpipe described in this work, they must be fixed in the cork that closes the passage through which the current of air arrives. $C\,c$ and $C'\,c$ (pl. 1, fig. 19) are two glass beaks, $c\,c$ are the corks, which can indifferently be adapted to c, in the wooden vice, by which the various parts of the blowpipe are connected when it is in action.

Of whatever material the beak may be made, its orifice must be perfectly round, and the *size* of the orifice, as we have before observed, must have a relation to the size of the wick which is to be used with it. You can ascertain the diameters of the orifices by inserting into them a little plate of brass, having the form of a long isoceles triangle, such as is represented by pl. 1, fig. 2. It should be an inch long, the twelfth of an inch wide at one end, and diminish to nothing at the other. When divided into eight equal parts, it will give, at the divisions, the respective proportions of 1, 2, 3, 4, 5, 6, 7 *eighths* of the diameter at the wide end, as is exemplified by the figure above referred to. We have stated in the following table the relative diameters which long experience has recommended to us, as being adapted to produce the greatest effect; yet it is not to be imagined that these proportions are mathematically correct and indispensable for the obtaining of good results. A sensible difference of effect would be perceived, however, were these proportions departed from in a notable manner.

Diameter of the wick, *in inches.*	Diameter of the orifice of the beak, *in parts of an inch.*	Height of the wick above the surface of the oil, *in inches.*
¼	96th	½
½	48th	½
1	24th	¾
1½	16th	1
2	12th	1¼

It must be mentioned, that this table has been formed from experiments made with a glass-blower's lamp of the ordinary construction; so that, with the new lamp with the hood, described in this work, it will not be necessary to employ wicks of so great a bulk, nor yet to elevate them so much above the level of the oil, in order to produce the same effect. Hence there will be a very considerable saving in oil.

The wicks of a quarter of an inch in diameter are only adapted for mineralogieal examinations, for soldering very fine metallic substances, and for working very small tubes. When the objects are of considerable bulk, it is in general necessary to have a flame sufficiently large to cover the whole instrument, or at least all the portion of the instrument which is operated upon at once. For working tubes, of which the sides are not more than the twelfth of an inch in thickness, you should have a wick at least as wide as the tube that is worked upon. The diameter of the lamp-wick usually employed is *one inch*; a wick of this size is sufficient for all the glass instruments which are in common use.

II.

PRELIMINARY NOTIONS OF THE ART.

THE FLAME.

It is only by long habitude, and a species of routine, that workmen come to know, not only the kind of flame which is most proper for each object they wish to make, but the exact point of the jet where they ought to expose their glass. By analysing the flame, upon the knowledge of which depends the success of the work, we can immediately obtain results, which, without that, could only be the fruit of long experience.

Flame is a gaseous matter, of which a portion is heated to the point of becoming luminous; its form depends upon the mode of its disengagement, and upon the force and direction of the current of air which either supports its combustion or acts upon it mechanically. (Pl. 1, fig. 1.)

The flame of a candle, burning freely in still air, presents in general the form of a pyramid, of which the base is supported on a hemisphere. It consists of four distinct parts: the immediate products of the decomposition of the combustible by the heat which is produced, occupy the centre, o, where they exist in the state of an obscure gaseous matter, circumscribed by a brilliant and very luminous envelope, s; the latter is nothing but the obscure matter itself, in the circumstances where, on coming into contact with the atmosphere, it combines with the oxygen which exists therein, and forms what is properly called *flame.*

The blueish light which characterises the inferior part of the flame, s, is produced by a current of cold air, which, passing from below *upwards*, hinders the combustion from taking place at the bottom of the flame, at the *same* temperature that exists in the parts of the flame not immediately subject to this influence.

Finally, on observing attentively, we perceive a fourth part, which is but slightly luminous, and exists as an envelope of all the other parts of the flame. The greatest thickness of this envelope corresponds with the summit of the flame. From this point it gradually becomes thinner, till it arrives at the lowest part of the blueish light, where it altogether disappears. It is in this last-described portion of the flame that the combustion of the gas is finished, and there it is that we find the seat of the most intense heat which the flame of the candle affords. If we compare the temperature of the different parts of the flame, we find that the *maximum* of heat forms a ring corresponding to the zone of insertion, A A; a point which is the limit of the superior extremity of the blueish light.

When the flame is acted upon by the blowpipe, it is subject to two principal modifications:—

1. If, by means of a blowpipe with a very fine orifice, you direct a current of air through the middle of the flame, you project a portion of the flame in the direction of the blast. The jet thus formed appears like a tongue of fire, blueish, cylindrical, straight, and very long; the current of air occupies its interior. This flame is enveloped on all sides by an almost invisible light, which, extending beyond the blue flame, forms a jet, A′ B, very little luminous, but possessing an extremely high temperature. It is at the point A′, which corresponds with the extremity of the blue flame, that the *maximum* of heat is found. The extreme point of the jet B possesses a less degree of heat. This flame is adapted for mineralogical assays, for soldering, for working enamels, and in general for all small objects.

2. When the orifice of the blowpipe is somewhat large, or when (the orifice being capillary) the current of air is very strong, or the beak is somewhat removed from the flame, the jet of fire, instead of being prolonged into a pointed tongue, is blown into a brush. It makes then a roaring noise, and spreads into an irregular figure, wherein the different parts of the flame are confounded beyond the possibility of discrimination. This flame is very proper for the working of glass, and particularly of glass tubes; it ought to be clear and very brilliant, and above all should not deposit soot upon cold bodies suddenly plunged into it. The *maximum* of temperature in this flame is not well marked; we can say, however, that in general it will be found at about two-thirds of the whole length of the jet. As this roaring flame contains a great quantity of carburetted hydrogen, and even of vapour of oil, escaped from combustion, it possesses a disoxidizing or reducing property in a very high degree.

PLACES FIT TO WORK IN.

Every place is adapted for a workshop, provided it is not too light and the air is tranquil. The light of the lamp enables one to work with more safely than daylight, which does not permit the dull-red colour of hot glass to be seen. Currents of cold air are to be avoided, because they occasion the fracture of glass exposed to them on coming out of the flame.

MEANS OF OBTAINING A GOOD FIRE.

The lamp should be firmly seated upon a steady and perfectly horizontal table, and should be kept continually full of oil. The oil which escapes during the operation, from the lamp into the tin-stand placed below it, should be taken up with a glass tube having a large bulb, and returned to the lamp.

When you set to work, the first thing you have to do is to examine the orifice of the beak. If it is closed, or altered in form, by adhering soot, you must carefully clean it, and open the canal by means of a needle or fine wire. In the next place, you freshen the wick by cutting it squarely, and carrying off with the scissars the parts which are carbonised. You then divide it into two principal bundles, such as C, K (pl. 1, fig. 21), which you separate sufficiently to permit a current of air, directed between the two, to touch their surfaces lightly, without being interrupted in its progress. By pushing the bundles more or less close to one another, and by snuffing them, you arrive at length at obtaining a convenient jet. It is a good plan to allow, between the two principal bundles and at their inferior part, a little portion of the wick to remain: you bend this down in the direction of the jet, and make it lie immediately beneath the current of air.

The wick must be prevented from touching the rim of the lamp, in order to avoid the running of the oil into the stand of the lamp. This is easily managed by means of a bent iron-wire, disposed as it is in the lamp described in this work; see pl. 1, fig. 23, where the wire is seen in an elevated position. When the wick is in the lamp, the wire is brought down round the wick and level with the surface of the lamp. A few drops of oil of turpentine, spread on the wick, makes it take fire immediately, over its whole extent, on the approach of an inflamed substance.

To obtain a good fire, it is necessary to place the lamp in such a position that the orifice of the blowpipe shall just touch the exterior part of the flame. The beak must not enter the flame, as it can then throw into the jet only an inconsiderable portion of the ignited matter. See pl. 1, fig. 20. On the other hand, if the lamp be too far away from the blowpipe, the flame becomes trembling, appears blueish, and possesses a very low degree of heat.

For mineralogical experiments, and for operations connected with watch-making and jewellery, the current of air should project the flame horizontally. For glass-blowing, the flame should be projected in the direction intimated by the arrow in pl. 1, fig. 20—that is to say, under an angle of twenty or twenty-five degrees.

The current of air ought to be constant, uniform, and sufficiently powerful to carry the flame in its direction. When it is not strong enough to produce this effect, it is necessary to add weights to the bellows or the bladder, according as the glass-blowers' table or our lamp is employed. The point to which you should apply, in the use of these instruments, is to enable yourself to produce a current of air so uniform in its course that the projected flame be without the least variation.

Finally, when you leave off working you should extinguish the flame, by cutting off the inflamed portion of the wick with the scissars. This has the double advantage of avoiding the production of a mass of smoke and of leaving the lamp in a fit state for another operation.

CHOICE AND PRESERVATION OF GLASS.

The only materials employed in the fabrication of the objects described in this Treatise, are tubes of common glass or of flint-glass. They can be had of all diameters, and of every variety of substance. They are commonly about three feet long, but some are found in commerce which are six feet in length. You should choose tubes that are very uniform—that is to say, straight and perfectly cylindrical, both inside and outside. A good tube should have the same diameter from one end to the other, and the sides or substance of the glass should be of equal thickness in every part. This is indispensable when the tubes are to have spherical bulbs blown upon them. We shall describe, in the article *Graduation*, the method of ascertaining whether or not a tube is uniform in the bore.

The substance of the glass should be perfectly clear, without bulbs, or specks, or stripes. The tubes are so much the more easy of use, as the glass of which they are made is the more homogeneous. Under this point of view, the white glass, known in commerce by the name of crystal or flint-glass, is preferable to common glass: it is more fusible, less fragile, and less liable to break under the alternations of heat and cold; but it is dearer and heavier, and has the serious disadvantage of becoming permanently black when exposed to a certain part of the flame. This is an effect, the causes and consequences of which will be explained in a subsequent chapter.

You must take care never to employ flint-glass for instruments which are to be submitted to the action of certain fluids—such as sulphuretted and phosphuretted hydrogen, and the hydro-sulphurets; for these compounds are capable of decomposing flint-glass, in consequence of its containing oxide of lead. In general, hard common glass is preferable to flint-glass for all instruments which are to be employed in chemistry. Flint-glass should only be used for ornamental objects, and for the barometers, thermometers, and other instruments employed in philosophical researches.

It sometimes happens that glass tubes lose their transparence and ductility, and suddenly become almost infusible, in the fire of the lamp: this effect takes place when they have been kept for some time in a melted state. It is then almost impossible to bring them back to their original condition; it can only be done by exposing them for a long time to an exceedingly high temperature. You can prevent this accident by working such kind of glass with considerable rapidity, and in a pretty brisk fire. There are tubes, however, which vitrify so promptly that it is only a person well versed in the art who can make good use of them. It is best not to employ such glass. But how can it be discriminated before-hand? It is experience, sooner than any characters capable of description, that will teach you how to make choice of good glass; nevertheless we have observed, that, among the glass tubes which occur in commerce, those possessing a very *white colour* manifest this bad quality most particularly. It may be observed, that, for tubes which are to have thin sides, this vitrifiable sort of glass is better than any other.

For certain philosophical instruments it is necessary to employ flat tubes. These are formed of flint-glass, are very small, and have a canal or bore, which, instead of being round, as in common tubes, has the form of a long and very flat oval. This disposition has the advantage of rendering more perceptible the column of liquid that may be introduced, and which in a round canal would scarcely be visible. In choosing this sort of tubes, carefully avoid those of which the canal is twisted, and not found to be in the same plane, in the whole length of the tube.

The tubes should be sorted, according to their sizes and qualities, and should be deposited in large drawers or on long shelves, in such a manner as to be equally supported through their whole extent. They should also be sheltered from dust and from moisture. If you cannot conveniently warehouse them in this manner, you should tie them up in parcels, and support them in a perpendicular position.

It is a very bad plan to place them in an inclined direction, or to support them by their extremities on wooden brackets, as it is the fashion to do in chemical laboratories; because, as the tubes are then supported only at certain points, they bend, in course of time, under the influence of their own weight, and contract a curvature which is extremely prejudicial in certain instruments, and which it is almost impossible to correct.

PREPARATION OF TUBES BEFORE HEATING THEM.

Before presenting a tube to the flame, you should clean it well both within and without, in order to remove all dust and humidity. If you neglect to take this precaution, you run the risk of cracking or staining the glass. When the diameter of the tube is too small to permit of your passing a plug of cloth or paper to clean its interior, you can accomplish the object by the introduction of water, which must, many times alternately, be sucked in and blown out, until the tube is deemed clean. One end of it must then be closed at the lamp, and it must be gradually exposed to a charcoal fire, where, by raising successively all parts of the tube to a sufficiently high temperature, you endeavour to volatilize and expel all the water it contains. In all cases you considerably facilitate the disengagement of moisture by renewing the air in the tube by means of a bottle of Indian-rubber fastened to the end of a long narrow tube, which you keep in the interior of the tube to be dried during the time that it is being heated. You can here advantageously substitute alcohol for water, as being much more volatile, and as dissolving greasy matters; but these methods of cleansing should only be employed for valuable objects, because it is extremely difficult fully to expel moisture from a tube wherein you have introduced water, and because alcohol is too expensive to be employed where there is no particular necessity.

When the tubes no longer contain dust, or moisture, you measure them, and mark the divisions according to the sort of work which you propose to execute.

METHOD OF PRESENTING TUBES TO THE FIRE, AND OF WORKING THEM THEREIN.

The two arms are supported on the front edge of the table, and the tube is held with the hands either above or below, according as it may be necessary to employ more or less force, more or less lightness. You ought, in general, to hold the tube *horizontally*, and in such a manner that its direction may be perpendicular to that of the flame. Yet, when you wish to heat at once a large portion of the tube, or to soften it so that it shall sink together in a particular manner, as in the operation of sealing, you will find it convenient to *incline* the tube, the direction of which, however, must always be such as to turn the heated part continually towards you.

We are about to give a general rule, upon the observance of which we cannot too strongly insist, as the success of almost every operation entirely depends upon it. The rule is, *never to present a tube to the flame without* CONTINUALLY TURNING *it*; and turning it, too, with such a degree of rapidity that every part of its circumference may be heated and softened to the same degree. As melted glass necessarily tends to descend, there is no method of preventing a heated tube from becoming deformed but that of continually turning it, so as to bring the softened part very frequently uppermost. When you heat a tube near the middle, the movement of the two hands must be *uniform* and *simultaneous*, or the tube will be twisted and spoiled.

When the tubes have thick sides, they must not be plunged *into* the flame until they have previously been strongly heated. You expose them at first to the current of hot air, at some inches from the extremity of the jet; you keep them there some time, taking care to turn them continually, and then you gradually bring them towards, and finally into, the flame. The thicker the sides of the tubes are, the greater precaution must be taken to elevate the temperature gradually: this is the only means of avoiding the fractures which occur when the glass is too rapidly heated. Though it is necessary to take so much care with large and thick tubes, there are, on the contrary, some tubes so small and so thin that the most sudden application of the fire is insufficient to break them. Practice soon teaches the rule which is to be followed with regard to tubes that come between these extremes.

Common glass ought to be fused at the *maximum* point of heat; but glass that contains oxides capable of being reduced at that temperature (such as flint-glass) require to be worked in that part of the flame which possesses the highest oxidating power. If you operate without taking this precaution, you run the risk of decomposing the glass. Thus, for example, in the case of flint-glass, you may reduce the oxide of lead, which is one of its constituents, to the state of metallic lead. The consequence of such a reduction is the production of a black and opaque stain upon the work, which can only be removed by exposing the glass, during a very long time, to the extremity of the jet.

You must invariably take the greatest care to keep the flame from passing into the interior of the tube; for when it gets there it deposits a greasy vapour, which is the ordinary cause of the dirt which accumulates in instruments that have been constructed without sufficient precaution as to this matter.

In order that you may not blacken your work, you should take care to snuff the wick of the lamp whenever you perceive the flame to deposit soot.

You can judge of the *consistence* of the tubes under operation as much by the *feel* as by the *look* of the glass. The degree of heat necessary to be applied to particular tubes, depends entirely upon the objects for which they are destined. As soon as the glass begins to feel soft, at a *brownish-red heat*, for example, you are at the temperature most favourable to good *bending*. But is it intended to *blow a bulb*? The glass must, in this case, be completely melted, and subjected to a full *reddish-white heat*. We shall take care, when speaking hereafter of the different operations to be performed, to mention the temperature at which *each* can be performed with most success.

When an instrument upon which you have been occupied is finished, you should remove it from the flame *gradually*, taking care to *turn* it continually, until the glass has acquired sufficient consistence to support its own weight without becoming deformed. Every instrument formed thus of glass requires to undergo a species of *annealing*, to enable it to be preserved and employed. To give the instrument this annealing, it is only necessary to remove it from the flame very gradually, allowing it to repose some time in each *cooler* place to which you successively remove it. The thicker or the more equal the sides of the glass, the more carefully it requires to be annealed. No instrument should be permitted to touch cold or wet bodies while it is warm.

III.

FUNDAMENTAL OPERATIONS IN GLASS-BLOW-ING.

All the modifications of shape and size which can be given to tubes in the construction of various instruments, are produced by a very small number of dissimilar operations. We have thought it best to unite the description of these operations in one article, both to avoid repetitions and to place those who are desirous to exercise this art in a state to proceed, without embarrassment, to the construction of any instrument of which they may be provided with a model or a drawing; for those who attend properly to the instructions given here, with respect to the fundamental operations of glass-blowing, will need no other instructions to enable them to succeed in the construction of all kinds of instruments capable of being made of tubes. These fundamental operations can be reduced to ten, which may be named as follows:—

1. Cutting.	6. Sealing.
2. Bordering.	7. Blowing.
3. Widening.	8. Piercing.
4. Drawing out.	9. Bending.
5. Choking.	10. Soldering.

We proceed to give a detailed account of these different operations.

1.—CUTTING.

The different methods of cutting of glass tubes which have been contrived, are all founded on two principles; one of these is the division of the surface of glass by cutting instruments, the other the effecting of the same object by a sudden change of temperature; and sometimes these two principles are combined in one process.

The first method consists in notching the tube, at the point where it is to be divided, with the edge of a file, or of a thin plate of hard steel, or with a diamond; after which, you press upon the two ends of the tube, as if to enlarge the notch, or, what is better, you give the tube a slight smart blow. This method is sufficient for the breaking of small tubes. Many glass-blowers habitually employ an agate, or a common flint, which they hold in one hand, while with the other they rub the tube over the sharp edge of the stone, taking the precaution of securing the tube by the help of the thumb. For tubes of a greater diameter, you can employ a fine iron wire stretched in a bow, or, still better, the glass-cutters' wheel; with either of these, assisted by a mixture of emery and water, you can cut a circular trace round a large tube, and then divide it with ease.

When the portion which is to be removed from a tube is so small that you cannot easily lay hold of it, you cut a notch with a file, and expose the notch to the point of the blowpipe flame: the cut then flies round the tube.

This brings us to the second method of cutting tubes—a method which has been modified in a great variety of ways. It is founded on the property possessed by vitrified matters, of breaking when exposed to a sudden change of temperature. Acting upon this principle, some artists apply to the tube, at the point where they desire to cut it, a band of fused glass. If the tube does not immediately separate into two pieces, they give it a slight smart blow on the extremity, or they drop a little water on the heated ring. Other glass-blowers make use of a piece of iron heated to redness, an angle or a corner of which they apply to the tube at the point where it is to be cut, and then, if the fracture is not at once effected by the action of the hot iron, they plunge the tube suddenly into cold water.

The two methods here described can be combined. After having made a notch with a file, or the edge of a flint, you introduce into it a little water, and bring close upon it the point of a very little tube previously heated to the melting point. This double application of heat and moisture obliges the notch to fly right round the tube.

When the object to be cut has a large diameter, and very thin sides—when it is such a vessel as a drinking-glass, a cup, or a gas tube—you may divide it with much neatness by proceeding as follows. After having well cleaned the vessel, both within and without, pour oil into it till it rises to the point, or very nearly to the point, where you desire to cut it. Place the vessel, so prepared, in an airy situation; then take a rod of iron, of about an inch in diameter, make the extremity brightly red-hot, and plunge it into the vessel until the extremity of the iron is half an inch below the surface of the oil: there is immediately formed a great quantity of very hot oil, which assembles in a thin stratum at the surface of the cold oil, and forms a circular crack where it touches the sides of the glass. If you take care to place the object in a horizontal position, and to plunge the hot iron without communicating much agitation to the oil, the parts so separated will be as neat and as uniform as you could desire them to be. By means of this method we have always perfectly succeeded in cutting very regular zones from ordinary glass.

The method which is described in some works, of cutting a tube by twisting round it a thread saturated with oil of turpentine, and then inflaming the thread, we have found to be unfit for objects which have thick sides.

Some persons employ cotton wicks dipped in sulphur. By the burning of these, the glass is strongly heated in a given line, or very narrow space, which is instantly cooled by a wet feather or a wet stick. So soon as a crack is produced, it can be led in any required direction by a red-hot iron, or an inflamed piece of charcoal.

Finally, you may cut small portions from glass tubes in a state of fusion, by means of common scissars.

2.—BORDERING.

To whatever use you may destine the tubes which you cut, they ought, almost always, to be bordered. If you merely desire that the edges shall not be sharp, you can smoothen them with the file, or, what is better, you can expose them to the flame of the lamp until they are rounded. If you fear the sinking in of the edges when they are in a softened state, you can hinder this by working in the interior of the tube a round rod of iron, such as pl. 1, fig. 5. The rod of iron should be one-sixth of an inch thick; one end of it should be filed to a conical point, and the other end be inserted into a thin, round, wooden handle. You will find it convenient to have a similar rod with a slight bend in the middle.

When you desire to make the edges of the tube project, bring the end to a soft state, then insert in it a metallic rod, and move it about in such a manner as to widen a little the opening. While the end of the tube is still soft, place it suddenly upon a horizontal surface, or press it by means of a very flat metallic plate. The object of this operation is to make the end of the tube flat and uniform. The metallic rod which you employ may be the same as we have described in the preceding paragraph. Instead of agitating the rod in the tube, you may hold it in a fixed oblique position, and turn the tube round with the other hand, taking care to press it continually and regularly against the rod. See pl. 1, fig. 6. Very small tubes can be bordered by approaching their extremities to a flame not acted upon by the blowpipe; particularly the flame of a spirit-lamp.

When the edges of a tube are to be rendered capable of suffering considerable pressure, you can very considerably augment their strength by soldering a rib or string of glass all round the end of the tube—see pl. 1, fig. 12. Holding the tube in the left hand, and the string of glass in the right, you expose them both at once to the flame. When their extremities are sufficiently softened, you attach the end of the rib of glass to the tube at a very short distance from its extremity; you then continue gradually to turn the tube, so as to cause the rib of glass to adhere to it, in proportion as it becomes softened. When the rib has made the entire circumference of the tube, you separate the surplus by suddenly darting a strong jet of fire upon the point where it should be divided; and you continue to expose the tube to the flame, always turning it round, until the ring of glass is fully incorporated with the glass it was applied to. You then remove the instrument from the flame, taking care to anneal it in so doing. During this operation you must take care to prevent the sinking together of the sides of the tube, by now and then turning the iron rod in its interior. It is a *red heat*, or a *brownish red heat*, that is best adapted to this operation.

3.—WIDENING.

When you desire to enlarge the diameter of the end of a tube, it is necessary, after having brought it to a soft state, to remove it from the flame, and to press the sides of the glass outwards by means of a large rod of iron with a conical point. The tube must be again heated, and again pressed with the conical iron rod, until the proper enlargement is effected. This operation is much the same as that of bordering a tube with projecting edges.

4.—DRAWING OUT.

You can *draw out* or *contract* a tube either in the middle or at the end. Let us in the first place consider that a tube is to be drawn out in the middle. If the tube is long, you support it with the right hand *below*, and the left hand *above*, by which means you secure the force that is necessary, as well as the position which is commodious, for turning it continually and uniformly in the flame. It must be kept in the jet till it has acquired a *cherry red heat*. You then remove it from the flame, and always continuing gently to turn it, you gradually separate the hands from each other, and draw the tube in a straight line. In this manner you produce a long thin tube in the centre of the original tube, which ought to exhibit two uniform cones where it joins the thin tube, and to have the points of these cones in the prolongation of the axis of the tube. See pl. 1, fig. 3.

To draw out a tube at its extremity, you heat the extremity till it is in fusion, and then remove it from the flame; you immediately seize this extremity with the pliers, and at the same time separate the two hands. The more rapidly this operation is performed, the glass being supposed to be well softened, the more capillary will the drawn-out point of the tube be rendered. Instead of pinching the fused end with the pliers, it is simpler to bring to it the end of a little auxiliary tube, which should be previously heated, to fuse the two together, and then to draw out the end of the original tube by means of the auxiliary tube—see pl. 1, fig. 4 and 11. In all cases, the smaller the portion of tube softened, the more abrupt is the part drawn out.

When you desire to draw out a point from the side of a tube, you must heat that portion alone, by holding it fixedly at the extremity of the jet of flame. When it is sufficiently softened, solder to it the end of an auxiliary tube, and then draw it out. Pl. 1, fig. 18, exhibits an example of a tube drawn out laterally. A *red heat*, or a *cherry red heat*, is best adapted to this operation.

5.—CHOKING.

We do not mean by *choking*, the closing or stopping of the tube, but simply a diminution of the interior passage, or bore. It is a sort of contraction. For examples, see pl. 2, fig. 15, 20, 29. You perform the operation by presenting to the flame a zone of the tube at the point where the contraction is to be effected. When the glass is softened, you draw out the tube, or push it together, according as you desire to produce a hollow in the surface of the tube, or to have the surface even, or to cause a ridge to rise above it. A *cherry red heat* is the proper temperature to employ.

6.—SEALING.

If the sides of the tube to be sealed are thin, and its diameter is small, it is sufficient to expose the end that you wish to close to the flame of the lamp. When the glass is softened it sinks of itself, in consequence of the rotatory motion given to it, towards the axis of the tube, and becomes rounded. The application of no instrument is necessary.

If the tube is of considerable diameter, or if the sides are thick, you must soften the end, and then, with a metallic rod or a flat pair of pliers, mould the sides to a hemisphere, by bringing the circumference towards the centre, and continuing to turn the tube in the flame, until the extremity is well sealed, and perfectly round. Examples of the figure are to be seen in pl. 2, fig. 3 and 5. Instead of this method, it is good, when the extremity is sufficiently softened, to employ an auxiliary tube, with the help of which you can abruptly draw out the point of the original tube, which becomes by that means cut and closed by the flame. In order that this part may be well rounded, you may, as soon as the tube is sealed, close the other extremity with a little wax, and continue to expose the sealed part to the flame, until it has assumed the form of a *drop of tallow*. See pl. 2, fig. 15. You can also seal in this fashion, by blowing, with precaution, in the open end of the tube, while the sealed end is in a softened state.

If you desire the sealed part to be flat, like pl. 3, fig. 30, you must press it, while it is soft, against a flat substance. If you wish it to be concave, like the bottom of a bottle, or pl. 3, fig. 2, you must suck air from the tube with the mouth; or, instead of that, force the softened end inwards with a metallic rod. You may also draw out the end till it be conical, as pl. 2, fig. 4, or terminate it with a little button, as pl. 2, fig. 6. In some cases the sealed end is bent laterally; in others it is twirled into a ring, having previously been drawn out and stopped in the bore. In short, the form given to the sealed end of a tube can be modified in an infinity of ways, according to the object for which the tube may be destined.

You should take care not to accumulate too much glass at the place of sealing. If you allow it to be too thick there, you run the risk of seeing it crack during the cooling. Some farther observations on sealing will be found at the article *Water Hammer*, in a subsequent section. The operation of sealing succeeds best at a *cherry-red heat*.

7.—BLOWING.

The construction of a great number of philosophical instruments requires that he who would make them should exercise himself in the art of blowing *bulbs* possessing a figure exactly spherical. This is one of the most difficult operations.

To blow a bulb at the extremity of a tube, you commence by sealing it; after which, you collect at the sealed extremity more or less glass, according to the size and the solidity which you desire to give to the bulb. When the end of the tube is made thick, completely sealed, and well rounded, you elevate the temperature to a *reddish white* heat, taking care to turn the tube continually and rapidly between your fingers. When the end is perfectly soft you remove it from the flame, and, holding the tube horizontally, you blow quickly with the mouth into the open end, without discontinuing for a single moment the movement of rotation. If the bulb does not by this operation acquire the necessary size, you soften it again in the flame, while under the action of which you turn it very rapidly, lest it should sink together at the sides, and become deformed. When it is sufficiently softened you introduce, in the same manner as before, a fresh quantity of air. It is of importance to observe that, if the tube be of a large diameter, it is necessary to contract the end by which you are to blow, in order that it may be turned round with facility while in the mouth.

When the bulb which you desire to make is to be somewhat large, it is necessary, after having sealed the tube, to soften it for the space of about half an inch from its extremity, and then, with the aid of a flat piece of metal, to press moderately and repeatedly on the softened portion, until the sides of the tube which are thus pressed upon, sink together, and acquire a certain degree of thickness. During this operation, however, you must take care to blow, now and then, into the tube, in order to retain a hollow space in the midst of the little mass of glass, and to hinder the bore of the tube from being closed up. When you have thus, at the expense of the length of the tube, accumulated at its extremity a quantity of glass sufficient to produce a bulb, you have nothing more to do than to heat the matter till it is raised to a temperature marked by a *reddish-white* colour, and then to expand it by blowing.

Instead of accumulating the glass thus, it is more expedient to blow on the tube a series of little bulbs close to one another (see pl. 1, fig. 8), and then, by heating the intervals, and blowing, to unite these little bulbs into a large one of convenient dimensions.

We have already observed, and we repeat here, that it is indispensably necessary to hold the glass *out* of the flame during the act of blowing. This is the only means of maintaining uniformity of temperature in the whole softened parts of the tube, without which it is impossible to produce bulbs with sides of equal thickness in all their extent.

When you desire to form a bulb at the extremity of a capillary tube, that is to say, of a tube which has a bore of very small diameter, such as the tubes which are commonly employed to form thermometers, it would be improper to blow it with the mouth; were you to do so, the vapour which would be introduced, having a great affinity for the glass, would soon obstruct the little canal, and present to the passage of the air a resistance, which, with the tubes of smallest interior diameter, would often be insurmountable. But, even when the tubes you employ have not so very small an internal diameter, you should still take care to avoid blowing with the mouth; because the introduction of moisture always injures fine instruments, and it is impossible to dry the interior of a capillary tube when once it has become wet. It is better to make use of a bottle of Indian rubber, which can be fixed on the open end of the tube by means of a cork with a hole bored through it. You press the bottle in the hand, taking care to hold the tube vertically, with the hot part *upwards*; if you were not to take this precaution, the bulb would be turned on one side, or would exhibit the form of a pear, because it is impossible, in this case, to give to the mass in fusion that rotatory motion which is necessary, when the tube is held horizontally, to the production of a globe perfectly spherical in its form, and with sides of equal thickness.

Whenever you blow into a tube you should keep the eye fixed on the dilating bulb, in order to be able to arrest the passage of air at the proper moment. If you were not to attend to this, you would run the risk of giving to the bulb too great an extension, by which the sides would be rendered so thin that it would be liable to be broken by the touch of the lightest bodies. This is the reason that, when you desire to obtain a large bulb, it is necessary to thicken the extremity of the tube, or to combine many small bulbs in one, that it may possess more solidity.

In general, when you blow a bulb with the mouth, it is better to introduce the air a little at a time, forcing in the small portions very rapidly one after the other; rather than to attempt to produce the whole expansion of the bulb at once: you are then more certain of being able to arrest the blowing at the proper time.

When you desire to produce a moderate expansion, either at the extremity or in any other part of a tube, you are enabled easily to effect it by the following process, which is founded on the property possessed by all bodies, and especially by fluids, of expanding when heated; a property which characterises air in a very high degree. After having sealed one end of the tube and drawn out the other, allow it to become cold, in order that it may be quite filled with air; close the end which has been drawn out, and prevent the air within the tube from communicating with that at its exterior; then gradually heat the part which you desire to have expanded, by turning it gently in the flame of a lamp. In a short time the softened matter is acted on by the tension of the air which is enclosed and heated in the interior of the tube; the glass expands, and produces a bulb or swelling more or less extensive, according as you expose the glass to a greater or lesser degree of heat.

To blow a bulb in the middle of a tube, it is sufficient to seal it at one of its extremities, to heat the part that you wish to inflate, and, when it is at a *cherry-red* heat, to blow in the tube, which must be held horizontally and turned with both hands, of which, for the sake of greater facility, the left may be held above and the right below.

If the bulb is to be large, the matter must previously be thickened or accumulated, or, instead of that, a series of small bulbs first produced, and these subsequently blown into a single larger bulb, as we have already mentioned. See pl. 1, fig. 8.

For some instruments, the tubes of which must be capillary, it is necessary to blow the bulbs separately, and then to solder them to the requisite adjuncts. The reason of this is, that it would be too difficult to produce, from a very fine tube, a bulb of sufficient size and solidity to answer the intended purpose.

You make choice of a tube which is not capillary, but of a sufficient diameter, very cylindrical, with equal sides, and tolerably substantial: it may generally be from the twentieth to the twelfth of an inch thick in the glass. You soften two zones in this tube, more or less near to each other, according to the bulk you desire to give to the bulb, and you draw out the melted part in points. The talent consists in *well-centering*—that is to say, in drawing out the melted tube in such a manner that the thin parts or points shall be situated exactly in the prolongation of the axis of the little portion of the original tube remaining between them. This operation is technically termed drawing *a cylinder between two points*. The tube so drawn out is exhibited by pl. 1, fig. 4. You cut these points at some distance from the central or thick part, and seal one end; you next completely soften the little thick tube and expand it into a bulb, by blowing with the precautions which have already been described. You must keep the glass in continual motion, if you desire to be successful in this experiment. Much rapidity of movement, and at the same time lightness of touch, are requisite in the operation here described. It is termed *blowing a bulb between two points*. Pl. 1, fig. 10, exhibits a bulb blown between two points.

To obtain a *round* bulb, you should hold the tube horizontally; to obtain a *flattened* bulb, you should hold it perpendicularly, with the fused extremity turned above; to obtain a *pear-shaped* bulb, you should hold the fused extremity downwards.

When you are working upon a bulb between two points, or in the middle of a tube, you should hold the tube horizontally, in the ordinary manner; but you are to push the softened portion together, or to draw it out, according as you desire to produce a ridge or a prolongation.

When you are at liberty to choose the point from which you are to blow, you should prefer, 1st, that where the moisture of the breath can be the least prejudicial to the instrument which is to be made; 2dly, that which brings the part which is to be expanded nearest to your eye; 3dly, that which presents the fewest difficulties in the execution. When bulbs are to be formed in complicated apparatus, it is good to reflect a little on the best means of effecting the object. It is easy to understand that contrivances which may appear very simple on paper, present difficulties in the practical execution which often call for considerable management.

8.—PIERCING.

You first seal the tube at one extremity, and then direct the point of the flame on the part which you desire to pierce. When the tube has acquired a *reddish-white* heat, you suddenly remove it from the flame, and forcibly blow into it. The softened portion of the tube gives way before the pressure of the air, and bursts into a hole. You expose the tube again to the flame, and border the edges of the hole.

It is scarcely necessary to observe, that, if it be a sealed extremity which you desire to pierce, it is necessary to turn the tube between the fingers while in the fire; but if, on the contrary, you desire to pierce a hole in the side of a tube, you should keep the glass in a fixed position, and direct the jet upon a single point.

If the side of the tube is thin, you may dispense with blowing. The tube is sealed and allowed to cool; then, accurately closing the open extremity with the finger, or a little wax, you expose to the jet the part which you desire to have pierced. When the glass is sufficiently softened, the air enclosed in the tube being expanded by the heat, and not finding at the softened part a sufficient resistance, bursts through the tube, and thus pierces a hole.

You may generally dispense with the sealing of the tube, by closing the ends with wax, or with the fingers.

There is still another method of performing this operation, which is very expeditious, and constantly succeeds with objects which have thin sides. You raise to a *reddish white* heat a little cylinder of glass, of the diameter of the hole that you desire to make, and you instantly apply it to the tube or globe, to which it will strongly adhere. You allow the whole to cool, and then give the auxiliary cylinder a sharp slight knock; the little cylinder drops off, and carries with it the portion of the tube to which it had adhered. On presenting the hole to a slight degree of heat, you remove the sharpness of its edges.

When you purpose to pierce a tube laterally, for the purpose of joining to it another tube, it is always best to pierce it by blowing many times, and only a little at a time, and with that view, to soften the glass but moderately. By this means the tube preserves more thickness, and is in a better state to support the subsequent operation of soldering.

There are circumstances in which you can pierce tubes by forcibly sucking the air out of them; and this method sometimes presents advantages that can be turned to good account. Finally, the orifices which are produced by cutting off the lateral point of a tube drawn out at the side, may also be reckoned as an operation belonging to this article.

9.—BENDING.

If the tube is narrow, and the sides are pretty thick, this operation presents no difficulty. You heat the tube, but not too much, lest it become deformed; a *reddish-brown* heat is sufficient, for at that temperature it gives way to the slightest effort you make to bend it. You should, as much as possible, avoid making the bend too abrupt. For this purpose, you heat a zone of one or two inches in extent at once, by moving the tube backwards and forwards in the flame, and you take care to bend it very gradually.

But if the tube is large, or its sides are thin, and you bend it without proper precautions, the force you employ entirely destroys its cylindrical form, and the bent part exhibits nothing but a double flattening,—a canal, more or less compressed. To avoid this deformity it is necessary, first, to seal the tube at one extremity, and then, while giving it a certain curvature, to blow cautiously by the other extremity, which for convenience sake should previously be drawn out. When tubes have been deformed by bad bending, as above described, you may, by following this method, correct the fault; that is to say, upon sealing one extremity of the deformed tube, heating the flattened part, and blowing into the other extremity, you can with care reproduce the round form.

In general, that a curvature may be well-made, it is necessary that the side of the tube which is to form the concave part be sufficiently softened by heat to sink of itself equally in every part during the operation, while the other side be only softened to such a degree as to enable it to give way under the force applied to bend it. On this account, after having softened in a *cherry-red heat* one side of the tube, you should turn the other side, which is to form the exterior of the curvature, towards you, and then, exposing it to the point of the jet, you should bend the tube immediately upon its beginning to sink under the heat.

When you desire to bend the extremity of a tube into a ring you must employ a metallic rod, with which, by pressing on the tube, you separate with a curve, C, (see pl. 1, fig. 14) all the portion A C which is necessary to produce the desired curl. You then successively soften all parts of this curve, and gradually twist it in the direction indicated by the arrow, pressing the iron rod constantly upon the extremity of the curve. When the end A comes into contact with bend C you solder them together at this point, and thus complete the ring. Pl. 2, fig. 27, and pl. 3, fig. 27, exhibit examples of rings formed by this process.

10.—SOLDERING.

If the tubes which you propose to solder are of a small diameter, pretty equal in size, and have thick sides, it is sufficient, before joining them together, to widen them equally at their extremities, by agitating a metallic rod within them. (Pl. 1, fig. 17.)

But if they have thin sides, or are of a large diameter, the bringing of their sides into juxta-position is very difficult, and the method of soldering just indicated becomes insufficient. In this case you are obliged to seal, and subsequently to pierce, the two ends which you desire to join. The disposition which this operation gives to their sides very much facilitates the soldering.

Finally, when the tubes are of a very different diameter, you must draw out the extremity of the larger and cut it where the part drawn out corresponds in diameter to the tube which it is to be joined to. Pl. 1, fig. 9 and 15, exhibit examples of this mode of adapting tubes to one another.

For lateral solderings you must dispose the tubes in such a manner that the sides of the orifices which you desire to join together coincide with each other completely. See pl. 1, fig. 7.

When the holes are well prepared, you heat at the same time the two parts that are to be soldered together, and join them at the moment when they enter into fusion. You must push them slightly together, and continue to heat successively all their points of contact; whereupon the two tubes soon unite perfectly. As it is almost always necessary, when you desire the soldering to be neatly done, or the joint to be imperceptible, to terminate the operation by blowing, it is proper to prepare the extreme ends of the tubes before-hand. That end of the tube by which you intend to blow should be carefully drawn out, provided it be so large as to render drawing out necessary; and the other end of the tube, if large, should be closed with wax, as in pl. 1, fig. 9, or if small, should be sealed at the lamp (pl. 1, fig. 15). When the points of junction are perfectly softened, and completely incorporated with each other, you introduce a little air into the tube, which produces a swelling at the joint. As soon as this has taken place, you must gently pull the two ends of the joined tube in different directions, by which means the swelled portion at the joint is brought down to the size of the other parts of the tube, so that the whole surface becomes continuous. The soldering is then finished.

To solder a bulb or a cylinder between two points, to the extremity of a capillary tube, you cut and seal one of the points at a short distance from the bulb (pl. 1, fig. 16), and at the moment when this extremity is in fusion you pierce it by blowing strongly at the other extremity. By this means the opening of the reservoir is terminated by edges very much widened, which facilitates considerably its being brought into juxta-position with the little tube. In order that the ends of the two tubes may be well incorporated the one with the other, you should keep the soldered joint for some time in the flame, and ought to blow in the tube, push the ends together and draw them asunder, until the protuberance is no longer perceptible.

If, after having joined two tubes, it should be found that there still exists an opening too considerable to be closed by simply pushing the two tubes upon one another, you can close such an opening by means of a morsel of glass, applied by presenting the fused end of an auxiliary tube.

You should avoid soldering together two different species of glass—for example, a tube of ordinary glass with a tube of flint-glass; because these two species of glass experience a different degree of contraction upon cooling, and, if joined together while in a fused state, are so violently pulled from one another as they become cool, that the cohesion of the point of soldering is infallibly overcome, and the tube breaks. You ought also, for a similar reason, to take care not to accumulate a greater mass of glass in one place than in another.

If the first operation has not been sufficient to complete the soldering, the tube must be again presented to the flame, and again pushed together at the joint, or drawn asunder, or blown into, according as it may appear to be necessary. In all cases the soldering is not truly solid, but inasmuch as the two masses of glass are well incorporated together, and present a surface continuous in all points.

The mineralogical flame (pl. 1, fig. 1, A′ B) is that which is to be employed in preference to the larger flame, when you desire to effect a good joining: it is sufficient to proportion the size of the flame to the object you wish to execute.

IV.

CONSTRUCTION OF CHEMICAL AND PHILOSOPH-ICAL INSTRUMENTS.

When a person is well acquainted with the fundamental operations which we have just described, the preparation of the instruments of which we are about to speak can present scarcely any difficulty. Indeed, some of them are so extremely simple, and are so easy of execution, that it is sufficient to cast a glance upon the figures which represent them, to seize at once the method which must be followed in their construction. Of such instruments we shall not stop to give a detailed description, but shall content ourselves with presenting the design.

On the other hand, it is of importance to observe that a certain number of instruments are *graduated* or furnished with pieces, or *mountings*, of which it is not the object of our art to teach the construction, and which demand a more or less extensive knowledge of the sciences. We shall treat of these mountings but summarily, referring the student, for more detailed instructions, to the works on natural philosophy and chemistry, in which these instruments are especially treated of. Our reason for this is, that we do not wish to abandon the plan we had adopted of describing simply the art of glass-blowing. To describe the use and application of philosophical instruments, or to explain the principles on which they act, would be passing quite out of our province.

ADAPTERS.

These are tubes of glass of various forms, employed in chemistry to connect together the different pieces constituting an apparatus—as, for example, to join a retort to a receiver during the operation of distillation. You should take care to border the extremities of an adapter; or you may widen them into the form of the mouth of a bottle, when they are to be closed air-tight by corks. Besides this, there is nothing particular to be observed in the preparation of adapters.

Apparatus for Boiling in Vacuo.—Represented by pl. 3, fig. 19. Employ a tube about a quarter of an inch in diameter. Blow two bulbs; give the tube the necessary curvature; fill one of the bulbs with nitric ether; boil the ether to expel the atmospheric air from the apparatus, then seal the opening in the other bulb.

CRYOPHORUS

Apparatus for Freezing in Vacuo. *The Cryophorus.*—Take a tube one-third of an inch or rather more in diameter, and pretty thick in the sides. Blow a bulb at each end; the first at the sealed part of the tube, the other at the open point; then give to the tube the curvature represented by pl. 3, fig. 32. Introduce as much water as will half fill one of the bulbs; make the water boil, and draw off the point and seal the apparatus during the ebullition.

APPARATUS FOR CONDUCTING WATER IN BENT TUBES.—Solder a funnel (see FUN-NELS) to the end of a tube; pierce two holes in this tube in the same line, and solder to each a little addition proper to receive a cork. Finish the instrument by bending it in the manner indicated by pl. 4, fig. 18.

APPARATUS FOR EXPERIMENTS ON RUNNING LIQUIDS.—A tube bent once at a right angle, mounted with a funnel, pierced laterally, and soldered at the same point to a smaller tube. See pl. 3, fig. 17.

APPARATUS FOR EXHIBITING THE PHENOMENA OF CAPILLARY TUBES.—This appa-ratus consists of a capillary tube soldered to another tube of a more considerable diameter. Sometimes it is bent like the letter U. Pl. 3, fig. 15.

APPARATUS FOR THE PREPARATION OF PHOSPHURET OF LIME.—An apparatus that can be employed for the preparation of phosphuret of lime, as well as in a variety of other chemical experiments, consists of a tube sealed at one extremity, slightly bent and choked at two inches and a half from the sealed part, and drawn out (af-ter the introduction of the substances to be operated upon) at the other extremity. This little distillatory apparatus is represented by pl. 3, fig. 29.

ARCHIMEDES'S SCREW.

There is no particular process for the making of this instrument. It is, howev-er, necessary for one who would succeed in making it, to exercise himself in the art of well bending a tube. After a few attempts, you may finish by producing a pretty-regular spiral. The tube chosen for this instrument should be six or seven feet long, and about one-third of an inch in diameter. You commence by making a bend, nearly at a right angle, about four inches from one of its extremities. This bent portion serves afterwards as a handle, and very much facilitates the opera-tion; it represents the prolongation of the rational axis which may be conceived to pass through the centre of the spiral. See pl. 4, fig. 10.

BARKER'S MILL.

Apparatus for exhibiting the rotatory motion produced by the running of liquids.—Contract a tube at its two extremities, pierce it laterally about the middle of its length, and solder to the hole an additional tube, terminated by a funnel. Soften the principal tube at the side opposite to the part that was pierced, and form there a conical cavity by pressing the softened glass inward with the aid of a metallic rod. This cavity must be so carefully made that the whole apparatus can be sup-ported on a pivot. Bend the contracted ends of the tube horizontally, and in differ-ent directions, cut off their extremities at a proper length, and slightly border the edges of the orifices. See pl. 3, fig. 33.

You may produce this apparatus under a different form, as may be seen at pl. 3, fig. 5.

BAROMETERS.

Barometers serve to measure the pressure of the atmosphere. The following are the varieties most in use.

Cistern Barometer.

Take a tube about thirty-two inches long, and at least one-third of an inch in diameter, internally; seal one of its extremities, free it with most particular care from moisture, fill it with mercury, and make the mercury boil in the tube, by heat, in order to drive out every particle of air which might be present. When the tube is full of mercury, and the boiling has taken place, turn it upside down, and plunge the open end into a cistern also filled with mercury which has been boiled. See pl. 2, fig. 4.

Dial (or Wheel) Barometer.

The tube intended for this barometer should be very regular in the bore. It should be thirty-nine inches long. Close it at one end, and bend it like the letter U at about thirty-two inches from the sealed extremity. See pl. 2, fig. 5, and *Graduation of the Dial Barometer*.

Syphon Barometer.

Make use of such a tube as might be employed for a *Cistern Barometer*; solder to its open end a cylindrical or spherical reservoir, and bend the tube close to the point of junction in such a manner as to bring the cylinder parallel with the tube. If the reservoir is to be closed with a cover of leather, cut off the remaining point of the cylinder, slightly widen the orifice, and then border it. If no leather is to be applied, but the point of the cylinder left open, it is necessary, after the introduction of the mercury, to draw off the point abruptly, and to leave an opening so small that mercury cannot pass by it. Pl. 2, fig. 6.

Stop-Cock Barometer.

This differs from the preceding barometer only by having a stop-cock mounted in iron between the reservoir and the tube.

Compound Barometers.

Blow a bulb at each end of a barometer tube of about thirty-three inches in length. Solder a small and almost capillary tube to the point which terminates one of the bulbs, and bend the great tube very near this bulb. This must be done in such a manner that the centre of one bulb shall be thirty inches from the centre of the other bulb. Introduce a quantity of mercury sufficient to fill the great tube and half the two bulbs; fill the remaining space in the last bulb with alcohol.

You may give a different disposition to this instrument. Divide a barometer tube into two, three, or four pieces, and reunite the pieces by intermediate capillary tubes, so as to form a series of large and small tubes, soldered alternately the one at the end of the other. Then communicate to this compound tube the form exhibited by pl. 3, fig. 25, and join, at each superior bend, a little tube, for the convenience of easily filling the instrument with mercury: seal these tubes as soon as the mercury is introduced. The graduation of compound barometers is made by bringing them into comparison with a good standard barometer. After taking two or three fixed points, it is easy to continue the scale.

Gay Lussac's Barometer.

Take a tube which is very regular in the bore, four-tenths of an inch in diameter, and thirty-five inches and a half in length. Seal one of its extremities and draw out the other; then cut the tube at about two-thirds of its whole length from the sealed end, and reunite the two pieces by means of a capillary tube soldered between them, the whole being kept in a line. See pl. 2, fig. 1. Pierce laterally the part of the tube which is drawn out, at some inches from the base of the point, and force the margin of the hole into the interior of the tube, by means of a conical point of metal, in such a manner as to form a little sunk funnel, of which the orifice must be very small. After having introduced the proper quantity of mercury into the instrument, boil it, and assist the disengagement of the bubbles of air by agitating a fine iron wire within the tube. Then remove the part of the tube which was drawn out, by sealing the end of the wide part. Give to the whole instrument the curvature indicated by pl. 2, fig. 3.

Bunten's Barometer.

This instrument differs from the preceding but in one point, namely, that the capillary tube is formed of two soldered pieces, of which the one, passing into the other, is terminated by a capillary point. This arrangement is exhibited by pl. 2, fig. 2.

Barometer pierced laterally for Demonstrations.

Take a tube thirty-nine inches long, with thick sides, and two-tenths of an inch internal diameter. Seal it at one end, and choke it at the distance of eight inches therefrom. Pierce a hole in the tube about twelve or sixteen inches from the choked part, and solder to the hole an additional piece, which can be closed by a cork or covered by a piece of bladder. The instrument is represented by pl. 2, fig. 15.

BELL-GLASSES FOR EXPERIMENTS.

These are pieces of tube sealed at one end, and widened or bordered at the other. They are extremely useful, and much employed in chemical experiments. They also supply the place of bottles for preserving small quantities of substances. Sometimes they are required to be straight, as pl. 3, fig. 12. Sometimes they need to be curved, as pl. 3, fig. 29. This is particularly the case when they are to be employed as retorts, for which purpose the sealed part should be made thin. Pl. 3, fig. 6, exhibits a retort with a tubulure.

BLOWPIPE.

We shall give in this article an account of the various pieces of glass which form part of the blowpipe described in the early part of this work. See pl. 1, fig. 19.

The beak C, which is employed with the candlestick, is merely a bent tube, at the extremity of which a bulb is blown. The bulb is terminated by a point, the thickness of the sides of which is augmented by turning it for a long time in the flame.

As for the beak used with the lamp, it is simply a bent tube C′, of which the orifice has been diminished by turning it round in the flame. The point of this beak is not drawn out like that of the beak described in the preceding paragraph, but is allowed to be thick, that it may not melt in the flame of the lamp.

The tube D F has four-tenths of an inch internal diameter, and is pretty thick in the sides. You must commence by bordering and slightly widening one of its extremities, and then proceed to choke it at about two inches from its other extremity, taking care to give to the choked part a figure as perfectly conical as possible, in order that the valve may act well. We have described the valve at length at p. 6.

The tube *d* is as much narrower than the tube D F as is necessary to permit it to pass up and down within the latter. Its use is to lengthen or shorten the tube for the convenience of the blower. The lower end is wound round with waxed thread, to make it fit air-tight. The mouth-piece is executed by widening the end of the tube, and then, while the widened part is still soft, by pressing the two sides obliquely, one against the other. By this means you give to the mouth-piece a flattened form, which adapts it better to the lips. The tube is finished by slightly bending this extremity.

In order that the bladder, or air reservoir, may be conveniently and securely attached to the tube E, you must take care to widen the end of this tube, and to turn up the edges strongly, by pressing the soft end against a flat metallic surface.

CAPSULES.

These are very small mercury funnels, of which the opening or neck has been closed. To transform these funnels into capsules, you must cut the neck as close as possible, and then soften, close, and flatten the opening. In performing this operation, hold the capsule by the edge with your pincers, and employ a piece of metal to press the glass together and make it close the hole and form the flat bottom of the capsule. See pl. 2, fig. 23.

Another Method.—After having blown a bulb at the end of a point, soften a narrow zone of the bulb, and then blow suddenly and strongly into it; by which means you separate the bulb into two capsules, which only need to be bordered. If you find any difficulty in presenting to the flame the capsule which forms the part of the bulb opposed to the point, you can attach to it a little rod of glass, which you can afterwards easily separate by a slight smart blow.

Occasionally you will have to make *capsules with double sides*, which will be described at the article *Nicholson's Hydrometer.*

CARTESIAN DEVILS.

Blow a bulb at the extremity of a very small tube, and heat a portion of the bulb, for the purpose of prolonging it into a beak. This can be effected with the aid of an auxiliary tube, which, on being joined to the heated part of the bulb, carries away with it the portion of glass which adheres. This portion of the bulb becomes thus prolonged into a little point, which must be cut at its extremity, so as to leave a small opening. The principal tube must be cut at the distance of half an inch from the bulb, and the ends of it must be drawn out and twisted into a ring. Instead of forming laterally a little beak to the bulb, you may pierce the tail, after twisting it into the form of a ring, or you may manage in such a manner as not to obliterate the canal of the twisted part. In general, little enamel figures are suspended to the ring of these globes, as is represented by pl. 2, fig. 22. A simple bulb, blown at the extremity of a small portion of tube, can supply the place of the Ludion or Cartesian devil. See pl. 2, fig. 8.

COMMUNICATING VASES.

Employ a tube of a large diameter; terminate one of its extremities with a funnel, fashion the other like the neck of a bottle; and bend the tube into the shape shewn by pl. 4, fig. 11. Then twist some other tubes into various forms, according to the end you propose to attain, and adjust these tubes to the neck of the large tube by means of corks, which have holes bored through them. In this manner an exchange of tubes is provided for various experiments.

DROPPING TUBES.

The name *dropping tube* is given to an instrument of glass which is very much employed in chemistry, for the purpose of transferring small quantities of liquor from one vessel into another, without disturbing either of the vessels. Dropping tubes are made of a great variety of forms and sizes, according to the purposes to which they are intended to be applied.

Blow a bulb between two points, and then, before the glass has regained its consistence, lengthen the bulb into an oval form. Cut and border the two points.

If the bulb, or reservoir, is to be so large that it cannot be formed at the expense of the thickness of the tube, and yet be sufficiently strong, it must be blown separately from a larger tube, and then soldered to two smaller tubes, one of which should have a certain curvature given to it. See pl. 2, fig. 20.

Sometimes a dropping tube is employed to measure small quantities of liquid. In this case the point should be drawn off abruptly, and the scale should be marked on the shank or tube with spots of black enamel.

Pl. 2, fig. 21, represents a peculiar variety of dropping tube employed in some experiments. It is made in the same manner as the common dropping tubes, excepting that, when the tail is formed, it is sealed at the extremity, bent there into a ring, and then pierced at A.

Pl. 3, fig. 26, represents another variety of dropping tube, a description of which is unnecessary.

FOUNTAINS.

It will readily be understood by those acquainted with the construction of hydraulic apparatus, that, by means of a judicious arrangement of glass tubes, a great variety of fountains may be produced. The following are given as examples.

Fountain of Circulation.

Take a tube, twenty-four or thirty inches long, nearly half an inch in diameter, and with pretty thick sides; blow a bulb at one of its extremities, and bend the other into a U, after having drawn it out as indicated by pl. 3, fig. 4. Pierce the tube at B, and join there a short piece adapted to receive a cork. Then prepare a bulb of the same size as the first bulb, and solder it to the extremity of a very long and almost capillary tube, which you must bend in zig-zag, in such a manner as to make it represent a Maltese cross, a star, a rose, or any other figure that may be suggested. The side of the bulb opposite to that which is attached to this twisted tube, ought to be formed like the neck of a bottle, in order that it may receive the drawn-out part of the larger tube, which should enter the bulb until the point of the large tube nearly touches the neck of the little tube at its junction with the bulb. This disposition is shewn in the figure. Seal now the other end of the little tube to the bulb of the large tube; then, with a little cement or sealing-wax, close the space between the bulb of the little tube and the point of the large tube. The instrument being thus prepared, as much alcohol, previously coloured red, must be inserted by the neck *b* as is sufficient to fill one of the bulbs. The neck is then closed with a cork, and a little cement or sealing-wax. Or, instead of forming this neck to the instrument, the additional piece may be drawn out to a point, which permits it to be sealed hermetically.

Fountain of Compression.

Introduce into a tube of large diameter a piece of capillary tube with thick sides. This must pass a little beyond the extremity of the large tube, which is to be softened and soldered to the other, so that it shall be fixed concentrically. The common point is then to be drawn out. When the tube is quite cold, and the small tube properly fixed in the centre of the large one, cut the latter at a proper distance, border it, and choke it near the end, which must be fashioned in such a manner as to be capable of being completely closed by a cork. See pl. 2, fig. 29.

Intermitting Fountain.

This apparatus is represented by pl. 3, fig. 16. Solder a cylindrical reservoir to the extremity of a capillary tube, pierced at *a*, and sealed at its extremity. Draw out abruptly the point of the reservoir, and give it a very small orifice; then give to the capillary tube the form indicated by the figure. Prepare next a funnel resembling a mercury-funnel, but much larger; choke the neck of this funnel, and bend the tube into the form of a syphon.

Hero's Fountain.

Solder a bulb to the extremity of a tube, and transform the bulb into a funnel. Close the funnel with a cork, and solder to the other end of the tube a bulb similar to the first. Next, solder a third bulb between two tubes, of which one must be twice as long the other; solder the longer of these tubes to the bulb of the first tube, and draw out the point of the shorter tube. You have now a long tube, with a funnel at one end, a contracted point at the other, and two bulbs in its length. Give to the whole apparatus the form indicated by pl. 3, fig. 21.

FUNNELS.

It will be seen, upon looking over the engravings, that funnels require to be made for a great variety of instruments; you ought therefore to acquire as soon as possible the art of making them well. The following are those most frequently required.

RETORT FUNNEL.—Blow a bulb at the extremity of a tube; present the superior hemisphere of the bulb to the flame, and when it is sufficiently softened, blow strongly into the other end of the tube. The air will force its way through the bulb, making a hole which will be larger or smaller according to the extent of surface which may have been softened. The opening of the funnel being made thus, there is nothing more to do than to adjust the edges, which, in the present state, are both fragile and irregular. This it is very easy to do. The edges are softened, the most prominent parts are cut off with the scissars, and the parts which are thin are bent back on themselves, that they may become thicker. Upon turning the funnel round in the flame, the smaller irregularities give way, and the edges become rounded. See pl. 2, fig. 24.

When the funnel is desired to be very large in proportion to the size of the tube, a bulb is made from a larger tube, and afterwards soldered to the small tube, and transformed into a funnel in the manner above described.

FUNNEL FOR INTRODUCING MERCURY INTO NARROW TUBES.—The mercury-funnel is represented by pl. 2, fig. 25. Blow a bulb between two points; cut off one of the points, and open the bulb at that place, in the manner described in the preceding article.

HYDROSTATIC FUNNEL.—This is represented by pl. 3, fig. 31. It is an instrument of constant use in chemical experiments. Form a funnel at the extremity of a tube in the manner described above, having previously blown a bulb near the middle of the tube. When this has been done, bend the tube into the form shown by the figure.

HOUR-GLASSES.

Blow four bulbs on a tube close to each other; open the two end bulbs like funnels, and then form them into flat supports or pedestals, according to the method described at the article *Test-glass with a foot*. Obstruct entirely the canal which separates one of these feet; choke to a certain extent the passage between the two remaining bulbs; and close the canal between the other foot and the bulbs, after introducing the quantity of sand which you have found to be necessary. See pl. 3, fig. 13.

HYDRAULIC RAM.

This instrument is represented by pl. 4, fig. 15. Employ a tube about six feet long, with thick sides and of large diameter. Seal it at one extremity, k, and border it at the other; solder at p an additional piece, choked so as to receive a valve. Pierce the tube at l; draw it out, and fix a funnel there; then twist the tube into a spiral. Form, on the other hand, a fountain of compression, o, and a funnel, m; and fix both of these pieces by means of sealing-wax, as soon as the two valves p and l have been put into their places.

HYDROMETERS.

Hydrometers are instruments which, on being plunged into liquids, indicate immediately their density or specific gravity. *Areometers* differ from hydrometers sometimes in graduation, sometimes merely in name. The following are examples of hydrometers, of which a great many varieties are in use.

Baumé's Hydrometer.

Make a cylinder between two points, and solder it to the extremity of a tube with thin sides, and which must be very regular on the outside. Close the open part which is to form the stalk of the hydrometer with a little wax. See pl. 1, fig. 9 and 15. When the soldering, which must be well done, is complete, and the stalk well centered, choke the reservoir at a little distance from the base of the point, by drawing it out in such a manner as considerably to diminish the canal in this part. Remove then the ball of wax which closed the tube, draw off the point of the cylinder, and make the part which was pulled away from the cylinder by the choking, into a bulb, by blowing with precaution into the tube. If the reservoir is required to be spherical instead of cylindrical, it must be softened and expanded by blowing. When it is intended to ballast the instrument with mercury, the canal must be completely stopped at the point where it is choked. In this case, the part drawn away from the cylinder is expanded into a bulb by blowing through the extreme point, which is to be cut off after the instrument is completed.

In the first case, you ballast the instrument with lead shot, which you fix in the lower bulb by means of a little wax, which closes the canal at the choked part. In the second case, after having proved the ballast by putting it first into the large reservoir, it is removed into the little bulb, and the latter is immediately sealed.

One of the essential conditions of a good hydrometer is that the stalk should keep a perfectly vertical position when the instrument is plunged in water. If, therefore, on proving the ballast, you perceive the stalk to rest obliquely, you must take care, on retiring it from the water, to wipe it dry, and to present the choked part between the cylinder and the little bulb to the flame; when it is softened, it is easy, by giving it a slight bend in the direction where the stalk of the hydrometer passes from the vertical, to rectify the defect.

Finally, when the instrument is ballasted, you must seal the stalk, after having fixed in its interior the strip of paper which bears the graduated division.

This method of operation serves equally for all the areometers known under the names of *areometer of Baumé*, *pèse-sels*, *pèse-liqueurs*, *pèse-acides*, and *hydrometers*, which differ only in the scheme of their graduation. As to the *size* and the *length* of the stalks, they depend upon the *dimensions* you desire to give to the degrees of the scale, and upon the *use* to which the instruments are destined. For the areometer of Baumé, and for the *pèse-sels*, the stalks are generally thicker and shorter than for hydrometers. Pl. 4, fig. 19, 20, and 21, represent different hydrometers.

Nicholson's Hydrometer.

Solder a bulb to the extremity of a capillary tube; open it so as to form a very wide funnel, or rather capsule; border the edges, and melt the point of junction with the tube so as to close the opening of the latter. Solder the other extremity of the tube to a cylindrical reservoir. Soften the point at the lower extremity of the cylinder, and obstruct the canal so as to convert the point into a glass rod; bend this rod into a hook. Now blow a bulb at the end of a point, as if to make a mercury funnel; but, after having softened the hemisphere of the bulb opposite to the point, and placed the latter in the mouth, instead of blowing into the bulb so as to make a funnel, strongly suck air from the bulb: by this means the softened part of the glass is drawn inwards, and you obtain a capsule with double sides, as exhibited by pl. 2, fig. 17. This capsule must have a small handle fastened across it, by which it may be hung to the hook formed at the bottom of the cylinder described above.

This hydrometer being always brought to the same level, the point to which it must be sunk in the liquid experimented with, is marked on the stalk by applying a little spot of black enamel. The instrument is represented by pl. 4, fig. 23. A variation in form is shewn by pl. 4, fig. 22.

Hydrometer with two Branches.

To measure the relative density of two liquids which have no action on each other, you employ a simple tube, bent in the middle and widened at its two extremities. See pl. 2, fig. 11.

Hydrometer with three Branches.

This consists of a tube bent in such a manner that the two branches become parallel. To this tube another is soldered at the point of curvature, and is bent in the direction exhibited by pl. 2, fig. 12. When the two branches are put into different liquids, and the operator sucks air from the third branch, the two liquids rise in their respective tubes to heights which are in the inverse ratio of their specific gravities.

Hydrometer with four Branches.

This is merely a tube bent three times, and widened at its extremities. Pl. 2, fig. 13.

To graduate hydrometers with two, three, and four branches, you have to divide their tubes into a certain number of equal parts.

MANOMETERS.

Make choice of a tube nearly capillary, very regular in the bore, and with sides more or less thick, according to the degree of pressure which it is to support. Seal this tube at one end, blow a bulb with thick sides near the middle, and curl it in S, just as is represented by pl. 2, fig. 9. For manometers which serve to measure the elasticity of the air under the receiver of the air-pump, what is generally employed is a tube closed at one end and bent into a U. Pl. 2, fig. 10. You should take care to contract these at some distance from the sealed part, in order to avoid the breaking of the instrument on the sudden admission of air. Manometers are graduated, as will be explained in the sequel.

MARIOTTE'S TUBE.

This is represented by pl. 2, fig. 7. It consists of a tube thirty-nine inches long, closed at one end, bordered and widened at the other, and bent into a U at the distance of eight inches from its sealed end. The graduation of this instrument will be described hereafter.

PHOSPHORIC FIRE-BOTTLE.

This is a short piece of tube closed at one end, and widened and bordered at the other, in such a manner as to receive a cork. Pl. 3, fig. 34. It is in this little vessel that the phosphorus is enclosed. Glasses of this form can be employed in a great variety of chemical experiments.

PULSOMETER.

This instrument consists of a tube, of which each extremity is terminated by a bulb; it is partly filled with nitric ether, and sealed at the moment when the ebullition of the ether has chased the atmospheric air wholly from the interior of the vessel. Pl. 2, fig. 16.

PUMP.

Solder a cylinder, B (pl. 4, fig. 12), to the extremity of a small tube, C, and form their point of coincidence into a funnel, to which you will adapt a valve. Pierce the wide tube or body of the pump at D, and solder there a piece of tube bent into an elbow and widened at the other end into a funnel, which is to be furnished with a second valve, as is represented in the figure. Prepare then the fountain of compression E, and, by means of a cork and a little sealing-wax, fix it upon the branch D. To prepare the piston, A, blow a bulb at the end of a tube, flatten the end of the bulb, and choke it across the middle, in order to form a place round which tow can be twisted, to make it fit the tube air-tight. Finish the piston by twisting the other end of the tube into a ring, as at A. The valves are formed of small cones of cork, or wood, having in the centre an iron wire of sufficient size and weight to enable them to play well.

RETORT FOR CHEMICAL EXPERIMENTS.

Plate 3, fig. 9, represents a combination of a large and a small tube, forming a retort, which can be employed with much advantage in many chemical experiments. When a gas is to be distilled by means of such a vessel, the ingredients are put into the wide tube, which is previously closed at one end, and then the other end of the tube is either drawn out or soldered to a narrow tube. Pl. 3, fig. 8 and 29, represent such vessels under different forms. Very often a sort of retort can be formed by joining a wide tube to a long bent narrow tube, by means of a cork.

TUBULATED RETORT.

This is represented by pl. 3, fig. 6. Prepare a retort, such as is described in the preceding article, but one which is bent near the closed end; pierce it at A (fig. 6), and solder there a little piece of tube previously drawn out and sealed, such as is represented by pl. 1, fig. 11. When the soldering is finished, soften the end of the little tube, pierce it, and fashion it into a bottle neck, so that it can be closed by a cork. Finish the instrument by forming the open end according to the purpose to which it may be destined. In the figure, the end is represented as drawn out for the convenience of blowing into the retort to pierce the tubulure.

THERMOSCOPE.

RUMFORD'S THERMOSCOPE.

This instrument is represented by pl. 3, fig. 35. It is necessary to take a tube almost capillary, to solder a bulb at each extremity, to pierce it laterally at *b*, and to solder there a piece of tube previously drawn out, but of which you open the point for the purpose of finishing the sealing of the bulb A. After doing this, you bend the two branches, as shewn in the figure. When the liquid has been introduced into the instrument, you must seal the little piece of tube which serves as a reservoir.

This instrument can be made in another manner. Take two pieces of tube, one of them twice as long as the other; solder a bulb at one end of each of these tubes, and at about the third part of the length of the long tube, parting from the bulb, bend it at a right angle; pierce the little tube at a corresponding distance, and solder to the hole the end of the long tube. The soldering being finished, and the whole system having the form indicated by pl. 3, fig. 35, introduce, by the open end of the short tube, a small quantity of coloured acid, and then seal the end of the short tube, which serves as a reservoir.

The interior diameter of the tubes which are generally employed as thermoscopes, is one-eighth or one-twelfth of an inch. The mode of graduation is described in a subsequent chapter.

SYPHONS.

The *simple syphon* is a glass tube bent, at a little distance from the middle, into a form which is intermediate between those of ∩ and ∧, the legs being stretched apart like those of the latter, but the bend being rounded like that of the former. The tube is bent *near* the middle, and *not exactly at* the middle, in order that the legs may be of unequal lengths; an arrangement which is indispensable. Syphons are made of different lengths and diameters, for various purposes. They can be made of tubes so capillary that it is sufficient to put them into water to make them act: the liquid rises in them by capillary attraction, and does not require to be sucked through the tube, as it does when large syphons are employed.

WIRTEMBERG SYPHON. — This syphon is the same as the simple syphon, excepting that the two branches are of equal length, and are bent in U at both extremities. Pl. 3, fig. 22.

SYPHON WITH THREE BRANCHES. — This instrument is represented by pl. 2, fig. 19. Close a tube at one end and draw it out at the other; pierce it at some inches from the contracted extremity, and solder to the hole a little tube of which the other end has been closed with wax. Give the tube the bend necessary to constitute a syphon, and open the two branches. The soldering of the two tubes is facilitated by giving to the extremity of the little tube a bend which adapts it to be applied parallel to the large tube. When the syphon is desired to be well finished, the mouth-piece of the little tube must be bordered and widened, and a bulb must be blown near the mouth-piece.

SYPHON WITH JET OF WATER. — This instrument is represented by pl. 3, fig. 1. Take a tube of a large diameter, close it at one end, and draw it out at the other. Cut the contracted part in such a manner as to be able to introduce, through the orifice, the extremity, also drawn out, of another tube, which should be almost capillary. Solder these together in such a manner that the point of the small tube shall remain fixed about an inch within the interior of the reservoir. Pierce again the latter, at B, and solder there another branch of the same diameter as the former; but fix it in such a manner that its side shall be contiguous to the side of the reservoir. Finally, give to the branches the bend represented by the figure.

SPOONS.

Solder a bulb to the extremity of a capillary tube; open the bulb as for a funnel, but make the opening laterally. Cut with scissars the edges of the part blown open, and in such a manner as to form a spoon or a ladle, according as the bulb had the form of a sphere or an olive. This instrument is useful for taking small quantities of acids. Pl. 3, fig. 11.

SPIRIT LEVEL.

The spirit level is represented by pl. 2, fig. 28. Choose a piece of tube very straight, and with sides precisely of the same thickness in all parts. Seal it at one end, and draw it out abruptly at the other. Fill it almost entirely with alcohol, and seal the point by the jet of a candle.

TEST GLASS WITH A FOOT.

Take a tube drawn out at one end; choke it at an inch from the base, in such a manner as to obstruct the canal almost entirely. Pl. 1, fig. 12. Cut off the point, close the opening, and soften the whole end completely; then blow it into a bulb and burst it into a funnel. Now present the contracted part to the fire, so as totally to close the passage. Border and soften the funnel, and by pressing it against a flat plate of metal give it the form of a foot, or pedestal. Cut the tube at the length which you desire the test-glass to have, and border the edges of the opening. This is a very useful little chemical instrument. It is represented by pl. 3, fig. 10.

THERMOMETERS.

Thermometers are instruments employed for appreciating changes of temperature, either in the atmosphere or in substances which we have occasion to examine. The following are the principal varieties now employed.

Ordinary Thermometer.

If you desire to make standard thermometers, you must have capillary tubes of perfect accuracy in the bore. You are assured of regularity in the diameter of a tube when a drop of mercury, made to pass along the canal by means of a gentle inclination, or by air blown from an Indian-rubber bottle, gives everywhere a metallic column of the same length.

For ordinary thermometers this precaution is superfluous. In all cases you employ a tube more or less capillary, at one of the extremities of which you blow or solder a spherical or cylindrical reservoir. See pl. 4, fig. 1 and 2. You fill the instrument with well-purified mercury, or alcohol, which you boil in the tube, in order to chase the air from it. As it is necessary to heat the instrument throughout its whole length, you must place it on a railing of iron wire, inclined in the manner represented by pl. 4, fig. 14, and covered with burning charcoal, or red-hot wood ashes. It is better, however, to employ a kind of muff, formed of two concentric wire grates, between which you put burning charcoal, and reserve the centre for the instrument. The tube is thus kept in a vertical position, which allows the bubbles of air to escape with more facility. An iron wire is made use of to fasten the tube precisely in the centre of the column of fire. The operation is considerably promoted by soldering a little funnel to the upper extremity of the thermometer tube; and, in order to avoid the interruption of the column of liquid by bubbles of air, it is better to give to the superior part of the reservoir the form of a cone (pl. 4, fig. 3), rather than to preserve the completely spherical form indicated by pl. 4, fig. 2.

When the ebullition has expelled all the air which was contained in the mercury, or alcohol, you immediately plunge the open extremity of the instrument into a vessel filled with one or the other of these liquids; or, instead of this, you pour the liquid into the funnel, in order that the instrument may be quite filled at the common temperature. You then cut off the funnel, if one has been used, and, by properly elevating the temperature of the reservoir, you expel so much of the liquid that the summit of the column rests at the point which you desire to make choice of for the mean temperature: this operation is termed *regulating the course of the thermometer*.

There are two methods of closing thermometers: you may either produce a vacuum above the column of mercury, or you may allow air to remain there. In the first case, after having drawn out the end of the tube, you heat the liquid until a single drop passes out of the opening; you then instantly bring the point into the jet, and seal it.

In the second case, you seal the instrument at the ordinary temperature, and having previously raised to a reddish-white heat the button of glass which is formed by the sealing, you suddenly elevate the temperature of the mercury. The liquid, on rising, compresses the enclosed air, which dilates the red-hot button at the summit of the tube, and produces a species of reservoir. This reservoir is indispensably necessary when you leave air above the column of liquid, in order to provide against the bursting of the instrument on those occasions when the temperature of the mercury comes to be considerably elevated. See pl. 4, fig. 13.

Dial Thermometer.

Terminate a piece of tube, of six-tenths of an inch in diameter, with two points, and solder to one of these points a tube one-eighth of an inch in diameter and six inches long; close the end of this small tube, and, heating a zone of the reservoir, near the base of the other point, blow a bulb there. Cut off the point by which you have blown, at a little distance from the bulb; open and border the end of the narrow tube, and bend it into a U. See pl. 4, fig. 16.

Fill the bulb and the reservoir with alcohol, and add a drop of mercury which fills a certain space in the narrow tube. This mercury bears on its surface a little iron weight, to which a thread is fastened; the other end of this thread passes over a pulley, whose axis turns a needle. The expansion or contraction of the alcohol causes the mercury to rise and fall, and consequently produces a movement of the needle or index of the dial. This thermometer is graduated like the others, by being brought into comparison with a standard thermometer.

Chemical Thermometer.

This instrument is merely a common thermometer, the divisions of which, graduated on paper, are enclosed in a very thin glass tube, to hinder them from being altered or destroyed when the instrument is plunged into liquids. Pl. 4, fig. 4, 5, 6, and 7, represent chemical thermometers of various kinds.

The case of the thermometer can be made in two different ways. According to the first, you take a tube of a pretty large diameter, and with very thin sides; you draw out one end and obliterate the point, which you bend into a ring, in a direction perpendicular to that of the case; you pass through this ring the stalk of the thermometer, which is thus placed parallel to the large tube. After having fixed the graduated scale in the interior of the case, by means of a small drop of sealing-wax, which has been dropped on the slip of paper, and which, being supported against the side of the case, needs only to be warmed to adhere there and fix the scale securely to its envelope, you close the upper extremity of the case by drawing it out, obliterating the canal and soldering it to the thermometer tube which has been introduced into the ring at the lower end of the case. You heat the connecting piece till it is soft, and then push the thermometer up and down until the zero marked on its tube corresponds with the zero marked on the scale within the case. See pl. 4, fig. 6 and 7.

The second method of making the case is as follows:—You take a tube with thin sides, and sufficiently large to contain the entire thermometer; you draw out the tube at one end, and choke it at some distance from the point of the contracted part. This you must do in such a manner as to form a little bulb, which is to be ballasted in the manner described at the article *Hydrometers*. After having introduced into the case a little ball of cotton, you place therein the thermometer, furnished with its scale, and in such a manner that the reservoir rests on the cotton. You terminate the upper end of the case either with a ring or by a contraction which permits the instrument to be suspended by a cord. See pl. 4, fig. 4 and 5.

Spiral Thermometer.

Take a tube which is not capillary, but which has thin sides; close one of its ends, and bend it round by pressing it with a metallic rod; continue to bend it round till it has made several turns, all in the same plane. See pl. 1, fig. 13. The latter turns may be managed with the fingers instead of the metallic rod. When the reservoir so formed is sufficiently large, solder to the end of it a capillary tube, which you point in a direction perpendicular to that of the axis of the spiral. The instrument is represented by pl. 4, fig. 8.

POCKET THERMOMETER.—The pocket thermometer differs in nothing from the thermometer just described, except that the capillary tube, instead of passing away from the spiral in a straight line, is turned round, so as to form a continuation of the spiral. See pl. 4, fig. 17.

Maximum Thermometer.

This instrument consists of an ordinary mercurial thermometer, bent at a right angle near the origin of the reservoir, and in the horizontal column of which a little steel or iron rod has been introduced: this rod, by gliding in the tube, where it experiences very little friction, serves as an index. Since this index does not permit the instrument to be sealed with the vacuum above the mercury, you must terminate the sealing by a little reservoir, as we have described at the article on the second method of closing thermometers. The instrument is represented by pl. 4, fig. 24.

Minimum Thermometer.

This instrument is constructed pretty nearly in the same manner as the preceding. The liquid, however, must be alcohol, and the index a little rod of enamel, which ought not to be quite so large as the bore of the thermometer tube. You seal the tube by making a vacuum above the column.

Bellani's Maximum Thermometer.

This thermometer is represented by pl. 4, fig. 9. Take a tube which is very regular, and about one-eighth or one-twelfth of an inch diameter in the bore; solder a reservoir at each end, one of them much larger than the other; make a bend near the large reservoir, and then fill the instrument with alcohol to A. Above that, place the first index, which consists of a very small piece of tube closed at one end and cut off square at the other. In the interior of this tube the two ends of a hair are fixed, by means of a little rod of iron, which is pushed into the tube. Introduce a quantity of mercury above this index, make the bend B, add again mercury as far as C, then another index similar to the first. Finally, fill the rest of the tube and the half the little reservoir with alcohol, and seal the point.

Differential Thermometer.

This instrument is represented by pl. 3, fig. 14. Take a tube ten or twelve inches long, and one-eighth or one-twelfth of an inch internal diameter; blow a bulb at one end, and bend the tube at a right angle towards the fourth part of its length. Prepare a second tube in the same manner, and solder the bent ends together, so as to form a single tube with a bulb at each end, having previously poured into one of the bulbs a small quantity of sulphuric acid tinged red.

Instead of following the above method, you may take a single tube of twenty or twenty-four inches in length, and of the above-mentioned diameter; you solder a bulb at each end, bend the tube twice till it represents the figure, pour in the acid, and then seal the open points. The graduation of the differential thermometer, as well as of all the other thermometers, is described in a subsequent section.

TUBES BENT FOR VARIOUS PURPOSES

Tube for Crystallizing Spermaceti.—Take a little capillary tube; curl one of its ends into a ring, and solder the other to a cylindrical reservoir, two-thirds of the capacity of which you fill with very pure spermaceti dissolved in sulphuric ether; you then seal the point of the reservoir. See pl. 3, fig. 27.

Tube for demonstrating the non-conductability of Heat by Liquids.—This is represented by pl. 2, fig. 26. It is a tube sealed at one end, bordered at the other, and bent in such a manner as conveniently to permit the upper part of a column of liquid to be exposed to heat.

Tube for estimating the Density of Vapours.—Represented by pl. 2, fig. 14. It is merely a tube sealed at one end, bordered at the other, and bent as shewn by the figure.

Tubes for exposing Substances to Heat and Gases.—This instrument consists of a tube bent in the middle into a U. Pl. 3, fig. 3. It is much employed in chemistry, for containing substances which we wish at the same time to expose to an elevated temperature and to the action of certain gases. This tube can also be employed for cooling gases, or liquids, in distillation; the bent part being, in this case, dipped into water or a freezing mixture, or enveloped in wet paper or cloth.

Tubes for the Preservation of Objects of Natural History, or of Chemical Preparations. — Take a tube of which the width and length corresponds with the object which is to be enclosed; draw it out at one end, and, after having obstructed the point, twist it into a ring. Introduce the object by the open extremity, which you must afterwards draw out; fill the tube with the liquid necessary to preserve the object, and then seal the point. See pl. 2, fig. 27.

If you desire to have the power of taking out the object at will — as, for example, when grain is preserved, or when, in chemistry, the tube is employed to contain salts and other compounds, of which small quantities are now and then required for use — you do not seal the end of the receiver, but border it in such a manner that it can be closed by a cork.

In some cases a cork is not sufficient to secure the substance from the action of air: it must then be assisted with a little cement. By melting together two parts of yellow wax, one part of turpentine, and a small quantity of Venetian red, a very useful cement for such purposes is obtained.

It is sometimes necessary to *suspend* the objects enclosed within the tube: you then introduce a little glass hook, the tail of which you solder to the upper extremity of the tube; managing this operation at the same time that you make the external ring for the support of the instrument. By turning the hook round cautiously, which is done when the end of the tube is in a soft state, and by cooling the whole with care, you may succeed in fixing the hook in the centre of the tube. See pl. 3, fig. 20.

Tube for emptying Eggs. — It is a simple tube, drawn out to a capillary point at one end, and bent there into a V. See pl. 3, fig. 23.

The application which the author has made of this instrument, and of the tube represented by pl. 3, fig. 26, has been shewn in a memoir inserted in the *Annales des Sciences Naturelles, Tom. XV. Novembre* 1828, concerning a new method of preparing and rendering durable collections of eggs destined for cabinets of Natural History.

VIAL OF THE FOUR ELEMENTS.

This instrument is represented by pl. 2, fig. 27. Take a tube drawn out at one end, obstruct the canal two inches from the extremity, and twist the contracted part into a ring. Draw out the other end of the tube, introduce the proper liquids, remove the point of the tube, and seal it. The liquids generally employed for filling the vial of the four elements are, 1. Mercury; 2. A very concentrated solution of carbonate of potash; 3. Oil of turpentine; 4. Alcohol. A portion of air is also allowed to remain in the tube.

WATER HAMMER.

Pl. 2, fig. 18, is a representation of this instrument. Choose a tube of a good diameter, and with thick sides; seal it at one end and draw it out at the other. Blow a bulb at the base of the contracted part; then, having put a quantity of water in the tube, let it boil therein, to expel the atmospherical air. When you imagine that all the air has been expelled, and that nothing remains in the tube but steam and water, seal the open point.

When you have to seal a tube in this manner, you should be careful to draw out the extremity of the tube somewhat abruptly, and leave a very small opening, so that it shall be sufficient to expose the point to the jet of a candle blown by a mouth blowpipe, to have the sealing completely and suddenly effected. You can afterwards round this sealed part by turning it in the flame of the lamp, provided, however, that you have preserved a sufficient thickness of glass at the sides of the point. If you omit to take this precaution, the pressure of the atmosphere, acting with great force on the softened glass when it is unsupported by the partial vacuum within the tube, is capable of producing such a flattening, or even sinking in of the matter, as could not subsequently be rectified; except, indeed, by heating simultaneously the liquid contained in the tube and the glass to be mended, which is an operation of a very delicate description.

WELTER'S SAFETY TUBES.

After having closed a tube at one end and drawn it out at the other, give it the curvature exhibited by plate 3, fig. 18. Pierce it then laterally, in the middle of the part *a b*, and solder there the extremity of a tube, to the other end of which a funnel has been soldered: it is necessary that the funnel be closed by a cork. The soldering being terminated, a bulb must be blown and the tube bent in S, in the manner shewn by the figure. Then open the closed end, and cut off the contracted point.

V.

GRADUATION OF CHEMICAL AND PHILOSOPHICAL INSTRUMENTS.

OF THE SUBSTANCES EMPLOYED IN THE PREPARATION OF THESE INSTRUMENTS.

Before proceeding to the subject of *graduation*, it is necessary to say a few words respecting the substances which are generally employed to fill a variety of instruments, particularly barometers and thermometers.

Mercury.—It ought to be completely purified from all foreign substances. You can separate it from the dust it may contain by passing it through a piece of chamois leather; you tie a very hard knot, and by pressure oblige the mercury to pass out in a fine rain. This process is sufficient for the purification of mercury which merely contains extraneous bodies in suspension; but it is not sufficient when the mercury to be purified contains tin, lead, or other metals, in solution. It is then necessary to distil the mercury; upon which the fixed metals remain behind. The oxide of mercury produced by the distillation is removed by agitating the distilled metal with sulphuric acid, and subsequently washing it with a large quantity of water, till all the acid is removed; it is then dried as completely as possible with blotting-paper, and afterwards is moderately warmed.

Alcohol ought to be very pure and well rectified. It is necessary to colour it, because, being colourless of itself, it could not be seen in capillary tubes. To colour alcohol, you infuse carmine in it, and, after some time, decant or filter the clear solution. The liquid should be perfectly transparent, and free from all extraneous substances. It is not proper to employ alcohol in the construction of standard thermometers; mercury being much preferable.

Sulphuric Acid.—It is made use of for the differential thermometer, and the thermoscope of Rumford. It has the advantage of being lighter than mercury, and very slightly volatile: these two qualities, joined to its tendency to absorb the vapour of water, render it very proper to be employed for various instruments. It must be very concentrated, and tinged red by carmine.

Ether.—Sulphuric and nitric ether, with which some small instruments are filled, are merely employed to shew with what facility these liquids are brought to their boiling point.

OF GRADUATION IN GENERAL.

Graduation, generally speaking, consists in dividing lines, surfaces, and capacities, into a certain number of equal or proportional parts. It is not our intention to treat here of the methods furnished by practical geometry for effecting such divisions with mathematical accuracy; these methods are known to every body. We shall confine ourselves to describing the processes of graduation which are peculiar to the instruments constructed by the glass-blower.

EXAMINATION OF THE BORE OF TUBES.

We have already observed, that, for standard thermometers and other instruments which require to be made very accurate, it is necessary to employ tubes which are extremely regular in the bore. When a drop of mercury, passed successively along all parts of the tube, forms everywhere a column of the same length, the examiner is assured of the goodness of the tube.

That a tube may be regular in the bore, it is not necessary that the bore be cylindrical; it is sufficiently accurate when equal lengths correspond to equal capacities. A tube with a flat canal, for example, can be perfectly accurate without at all approaching the cylindrical form. It is only necessary that a drop of mercury occupy everywhere the same length. We may observe, by,the way, that, in flat canals, the flattening should be always in the same plane.

DIVISION OF CAPILLARY TUBES INTO PARTS OF EQUAL CAPACITY.

As it is very difficult to meet with capillary tubes which are exactly regular in the bore, it happens that the tubes which glass-blowers are obliged to employ have different capacities in parts of equal length. You commence the division of these tubes into parts of equal capacity by a process described by M. Gay-Lussac. You introduce a quantity of mercury, sufficient to fill rather more than half the tube, and make a mark at the extremity of the column. You then pass the mercury to the other end of the tube, and again mark the extremity of the column. If you so manage that the distance between the two marks is very small, you may consider the enclosed space as concentric, and a mark made in the middle of the division will divide the tube into two parts of evidently equal capacity. You divide one of these parts, by the same process, into two equal capacities, and each of these into two others; and in this manner you continue to graduate the tube until you have pushed the division as far as you judge proper.

But it is still more simple to introduce a drop of mercury into the tube, so as to form a little cylinder, and then to mark the two extremities of the cylinder. If it were possible to push the drop of mercury from one end of the tube to the other, in such a manner as to make it coincide, at every removal, with the last mark, it would be very easy to divide the tube accurately; but as it is very difficult, not to say impossible, to attain this precision of result in moving the column of mercury, you must endeavour to approach exactness as nigh as may be. You measure, every time you move the mercury, the length of the cylinder it produces, and carry this length to the last mark, presuming the small space which is found between the mark and the commencement of the column to be fairly represented by the same space after the column. You thus obtain a series of small and corresponding capacities.

GRADUATION OF GAS JARS, TEST TUBES, &C.

If the tube is regular in the bore, close one end, either by sealing it at the lamp, or by inserting a cork, and pour into the interior two or three small and equal portions of mercury, in order to have an opportunity of observing the irregularities produced by the sealed part. Take care to mark, with a writing diamond, the height of the mercury, after the addition of each portion. When equal portions of mercury are perceived to fill equal spaces, take with the compass the length of the last portion, and mark it successively along the side of the tube, where you must previously trace a line parallel to its axis.

For tubes which are irregular in the bore, and where equal lengths indicate unequal capacities, it is necessary to continue the graduation in the same manner that you commenced it—that is to say, to fill the tubes by adding successively many small and equal portions of mercury, and marking the height of the metallic column after every addition. These divisions will of course represent parts of an ounce or of a cubic inch according to the measure which you make use of. When you have thus traced on the tube a certain number of equal parts, you can, by means of the compasses, divide each of them into two other parts of equal length. The first divisions being very close to one another, the small portion of tube between every two may be considered without much risk of error as being sensibly of equal diameter in its whole extent.

When the tube which you desire to graduate is long and has thin sides, it would be difficult to fill it with mercury without running the risk of seeing it break under the weight of the metal. In this case, you must use water instead of mercury.

Bell-glasses of large dimensions are graduated by filling them with water, placing them in an inverted position on a smooth and horizontal surface, which is slightly covered with water, and passing under them a series of equal measures of air. But it is then necessary to operate constantly at the same temperature and under the same atmospheric pressure, because air is very elastic and capable of being greatly expanded.

In all cases, tubes, bell-glasses, &c. ought to be held in a position perfectly vertical. The most convenient measure is a dropping-tube, on the stalk of which a mark has been made, or a small piece of tube, sealed at one end, and ground flat at the other; the latter can be accurately closed by a plate of glass.

The marks which are traced on tubes being generally very close to one another, you facilitate the reading of the scale by giving a greater length to those marks which represent every fifth division, and by writing the figures merely to every tenth division. See pl. 4, fig. 8. The number of divisions is somewhat arbitrary; nevertheless, 100, 120, 360, 1000, are divisions which, in practice, offer most advantages.

GRADUATION OF HYDROMETERS.

Cut a band of paper on which the graduation of the instrument can be traced, and let fall upon it a little drop of sealing-wax; then roll the paper upon a little glass tube, and introduce it into the stalk of the hydrometer. The instrument is afterwards to be plunged into distilled water, which is carefully kept at the temperature of 40° F. above zero. Give the instrument sufficient ballast to make it sink till the point (*a*, pl. 4, fig. 20,) which you desire to make to represent the density of water, touches the surface of the water. Mark this point with much precision; it is the zero of the instrument. The other degrees are taken by plunging the hydrometer into distilled water to which you have added 1, 2, 3, 4, 5, &c. *tenths*, or 1, 2, 3, 4, 5, &c. *hundredths*, of the substance for which you wish to construct the hydrometer, according as you desire the scale to indicate tenths or hundredths.

When you have thus marked the degrees on the stalk of the instrument, transfer them to the paper with the help of the compasses. The scale being completed, replace it in the tube of the hydrometer, where it must be fixed; in so doing, take care to make the degrees on the scale coincide precisely with those marked on the stalk.

You can thus procure hydrometers for alcohol, acids, salts, &c. which are instruments that indicate the *proportion* of alcohol, acid, salt, &c. contained in a given mass of water.

But if it were necessary to plunge the hydrometer in a hundred different solutions in order to produce the scale, it is easy to conceive that that would be extremely troublesome, especially for hydrometers which are employed in commerce, and which do not need to be so extremely accurate. When the density of the mixtures or solutions is a mean between those of the substances which enter into them, you may content yourself with marking the zero and one other fixed point, (*a* and *b*, pl. 4, fig. 20.) Then, as the stalk of the hydrometer is evidently of equal diameter in all its extent, you can divide the space which separates the two fixed points into a certain number of equal parts. One of these, being taken for unity, represents a particular quantity of the substance which you have added to a determined weight of distilled water. By means of this unity you can carry the scale up and down the stalk of the instrument. It is thus, that, to obtain a Baumé's hydrometer, after having obtained the zero by immersion in distilled water, you plunge the instrument into a solution containing a hundred parts of water and fifteen of common salt, to have the 15th degree, or containing a hundred water and thirty salt, to have the 30th degree. Upon dividing the interval into fifteen or thirty equal parts, according as you have employed one or the other solution, you obtain the value of the degree, which you can carry upwards or downwards as far as you wish.

Among the substances for which hydrometers are required in commerce, are some which it is impossible to obtain free from water—such are alcohol, the acids, &c. In this case it is necessary to employ the substances in their purest state, and deprived of as much water as possible.

The employment of hydrometers is very extensive: they are used to estimate the strength of lyes, of soap solutions, of wines, milk, &c. There is, in short, no branch of commerce in which these instruments are not required for the purpose of ascertaining the goodness of the articles which are bought and sold. The employment of hydrometers would be still more general, if they could be made to give immediately the absolute specific gravity of the liquids into which they might be plunged, the specific gravity of water being considered as unity. It is possible to graduate a thermometer of this description by proceeding as follows:—

Make choice of a hydrometer of which the exterior part of the stalk is very regular. Introduce the band of paper on which the scale is to be written, and then ballast the instrument. Make a mark where the surface of the distilled water touches the stalk. Remove the hydrometer from the water, wipe it perfectly dry, and weigh it very accurately with a sensible balance. Then pour into it a quantity of mercury equal to its own weight; plunge it again into the water, and again mark the point where the stalk touches the surface of the water. Pour the mercury out of the instrument, transfer the two marks to the scale, and divide this fixed distance into fifty equal parts. Having by this operation obtained the value of the degree, you carry it upwards and downwards, to augment the scale. If you take the first point near the reservoir, the hydrometer will be proper to indicate the density of liquids which are heavier than water; if you take it towards the middle of the tube, the contrary will be the case.

If you destine the hydrometer for liquids much heavier than water—such as acids, for example—you might, after having determined the first point, add to the original ballast as much mercury as is equal to the weight of the whole instrument; then the point where the stalk would touch the surface of the water, and which would be represented by 100, would be very high, and the second point, which would be found below, would be represented by 200. On dividing the space into a hundred equal parts, you would have the value of the degree, which could be carried up and down for the extension of the scale.

The specific gravities being in the inverse ratio of the volumes plunged into the liquid, the numbers of the scale which mark the specific gravities diminish from below; so that, on marking the lowest point 100, you have, on proceeding upwards, the successive degrees 0·99, 0·98, 0·97, 0·96, &c.

The hydrometers with two, three, and four branches, are graduated by having their tubes divided into a hundred or a thousand equal parts. The divisions on each branch must correspond with those on the other branches.

GRADUATION OF BAROMETERS.

The graduation of this instrument consists in dividing a piece of metal, wood, or ivory, into inches and parts of inches. The divided rod is then employed to measure the height of the mercury in the tube. As the rule is moveable, the operation presents no sort of difficulty: all that is necessary is to make the zero of the scale coincide with the inferior level of the mercury; the point which corresponds with the superior level of the mercury, seen in the tube, indicates the height of the barometric column. It is in this manner that the cistern barometer is graduated.

But if the barometer is one of those in which the surface of the mercury is variable, such as the barometer of Gay-Lussac, it is necessary to have recourse to a different process of graduation. If the two branches of the instrument are very regular, and of equal diameter, you first measure with precision the height of the column of mercury, then divide it in the middle, and fix the scale, which must be graduated in such a manner that the mark of fifteen inches corresponds exactly with the middle point. This mode of graduation serves to indicate merely the apparent height of the barometric column. If you desire that the scale should immediately indicate the real height, you must fix the zero at the middle of the column, and then double the figure which marks each degree.

When you do not wish to write the real height, you make two divisions, of which one proceeds upwards, the other downwards. You do not, in this case, double the value of each division, but in observations made with such a barometer scale you add the degree marked by the two surfaces, in order to find the real height.

It is in an analogous manner that you graduate the gauges or short barometers which are employed to measure the density of air under the recipient of the air-pump. You take the height of the mercury in the gauge, and fix at the middle of the column the zero of a double scale, of which one division proceeds upwards, the other downwards; or, instead of this, if you choose to have only one scale, and that an ascending scale, you double the value of every degree.

The zero of the barometric scale can be fixed below the inferior surface of the mercury; but then, to have the real height, it is necessary to measure precisely the height of the mercury in the two branches of the instrument, and to deduct the smaller from the larger.

Dial (or Wheel) Barometer.—The disposition which should be given to this instrument is precisely the same as that of the *Dial Thermometer*, described in a preceding section. You make a small iron weight float on the inferior surface of the mercury, and fix to this weight a silk thread, which is stretched by a counterpoise, and rolls over a very moveable pulley. The axis of this pulley carries a needle, which turns backwards or forwards according as the column of mercury augments or diminishes. You arrange the whole in such a manner that the extreme variations of this column cannot make the needle describe more than one circumference; with this view you give the pulley a diameter of nearly an inch.

The dial barometer being rather an object of luxury than an instrument of precision, you graduate it by inscribing the following words, at full length, on the scale. In pl. 4, fig. 16, for example, you write,

At the point	a	......	Tempest.
...	b	......	Much rain.
...	c	......	Rain or Wind.
...	d	......	Temperate.
...	e	......	Fine Weather.

...	f	......	Fixed Fair.
...	g	......	Very Dry.

You write nothing at the inferior division.

GRADUATION OF THE MANOMETER.

The graduation of this instrument consists in dividing the tube where the air is to be compressed, into a given number of parts of equal capacity; but as, in general, such tubes are employed as are nearly capillary and very regular, the operation is reduced to a linear division, where every degree occupies an equal space.

GRADUATION OF THERMOMETERS.

Construction of Standard Thermometers.—Having constructed your instrument with a very regular tube, or one which has been divided into parts of equal capacity, and having filled it with the proper liquid, according to the instructions given in a preceding section, the graduation is to be effected as follows. Procure very pure ice, break it into small pieces, and fill a vessel with it. When the ice begins to melt, plunge the thermometer into the middle of it, in such a manner that, without touching the sides of the vessel, the whole thermometer, or at least that part of it which contains the liquid, may be covered with ice. Allow the instrument to remain in this state until, in spite of the gradual melting of the ice, the surface of the column of liquid remains at a fixed point, and neither falls nor rises. Mark this point very carefully on the stalk of the thermometer, either with a thread or a little drop of sealing-wax, or with the trace of a diamond or a flint. This is the *freezing point*, the *zero* of the centigrade scale, the thirty-second degree of Fahrenheit's scale.

As for the second fixed point, it is marked during an experiment with boiling water, performed as follows:—You employ a vessel of tin plate sufficiently high to enclose the whole thermometer; you pour into this vessel distilled water, till it is about an inch deep, and then you heat it. The vessel is surmounted by a cover pierced with two holes, one of which is intended to receive the stalk of the thermometer, the other to allow the steam to escape. When, on continuing the ebullition, you observe that the mercury ceases to rise in the tube, you mark the point at which it has stopped, just as you marked the first point. The last mark indicates the *boiling point*; the one hundredth degree of the centigrade scale, the two hundred and twelfth degree of Fahrenheit's scale. You transfer to paper the distance which is found between the first point and the second point determined, and you divide this distance into one hundred equal parts, or degrees, for the centigrade thermometer, into eighty parts for the thermometer of Réaumur, and into one hundred and eighty for that of Fahrenheit. If the tube of the instrument is very regular in the bore, the degrees should be equal in length; if, on the contrary, you have been obliged to divide it into parts of equal capacity, you find how many of these parts or little spaces it is necessary to take to constitute one of the above degrees. You find this by dividing their whole number by 100, or 80, or 180, according to the degrees of the scale which you intend to make use of. Thus, if you find between the two points fixed by melting ice and boiling water, three hundred divisions of equal capacity, it is necessary to include *three* of these divisions in every *degree* of the centigrade scale.

The vessel employed to take the boiling point must be of metal, and its surface should be perfectly clean and well polished, and have no rough points. If sand, or other matters, were permitted to repose on the vessel, and to form asperities, the water would enter into ebullition at an inferior temperature.

This operation should, moreover, be performed under an atmospherical pressure, which is indicated by the barometer when the mercury stands at twenty-nine inches and a half. But as this pressure is different according to the elevation of the place of operation, and, indeed, suffers continual variations even in the same place, it follows that the temperature of boiling water is subject to continual changes, and that, in the graduation of the thermometer, it is indispensably necessary to take notice of the height of the barometer at the very moment that the point denoting the degree of boiling-water is fixed upon. You succeed in making the necessary corrections by the help of the following table, which is founded on the experiments of Sir G. Shuckburg and of the Committee of the Royal Society.

[See the Table on the opposite page.]

Common Thermometers. — Having, by the method which we have just described, obtained a *Standard Thermometer*, you may procure with facility as many ordinary thermometers as you desire. It is proper to employ the most regular tubes which you can obtain, and when the instruments are ready to be graduated, you must bring them into comparison with your standard thermometer. You place them together into a liquid of which you gradually raise the temperature, and you mark several points on the scale of the new thermometer, the intervals between which are subsequently divided into as many degrees as are marked on the scale of the standard thermometer. Thus, for example, you mark the 10° and 15°, and afterwards divide the interval into five equal parts. This gives you the length of a degree on the stalk of the new instrument. The more you multiply these fixed points, the more you insure the precision of the thermometer. When you have taken a certain number of points, you measure the remainder with the compasses.

Height of the Barometer in Inches.		Correction in 1000ths of the interval between the freezing and boiling points of Water.	
When the boiling point is found by immersing the Instrument in Steam.	When the boiling point is found by immersing the Instrument in Water.		
...	30.60	10	
...	30.50	9	
30.71	30.41	8	
30.50	30.29	7	
30.48	30.18	6	
30.37	30.07	5	Lower.
30.25	30.95	4	
30.14	30.84	3	
30.03	30.73	2	
29.91	30.61	1	
29.80	30.50	0	
29.69	29.39	1	
29.58	29.28	2	
29.47	29.17	3	
29.36	29.06	4	
29.25	28.95	5	
29.14	28.84	6	Higher.
29.03	28.73	7	
28.92	28.62	8	
28.81	28.51	9	
28.70		10	
		The boiling point to be marked so much higher or lower than the stand of the mercury during the experiment.	

The zero, 0°, of the thermometer of Fahrenheit, is taken by means of a mixture of snow and common salt, and its maximum point is, like that of the preceding thermometer, taken by means of boiling water; but this interval is divided into 212 degrees; so that the scale marks 32° where the centigrade and Réaumur's scales mark 0°.

The thermometer of Delisle has but one fixed point, which is the heat of boiling water; this is the zero of the instrument. The inferior degrees are 0,0001 (one ten-thousandth part) of the capacity of the bulb and stalk of the thermometer. It marks 150° at 0° of the centigrade, or 32° of Fahrenheit's thermometer.

The dial, the maximum and the minimum thermometers, are graduated according to the same principles as the common thermometers.

You can, with a mercurial thermometer, make the centigrade scale rise to 300 or 400 degrees above zero; but with an alcohol thermometer, you must never go beyond the heat of boiling water. On the contrary, the inferior degrees of the alcohol thermometer can be carried to the very lowest point, while those of the mercurial thermometer should be stopped at thirty or thirty-five degrees below the zero of the centigrade scale, as the mercury then approaches very near the point of its congelation. In all cases, the degrees of thermometer scales are indicated by the sign - when they are below zero, and by the sign + when they are above it; the - is always marked, but the + generally omitted. See pl. 4, fig. 6.

We may observe here that it is proper from time to time to plunge the standard thermometer into melting ice, for the purpose of verifying its exactness. It has been found that thermometers constructed with a vacuum above the column of mercury gradually become inaccurate, the 0° ascending, until it corresponds with + 1° or + 2°. This singular effect is attributable to the constant pressure of the atmosphere, which, being supported merely by the resistance of the very thin sides of the thermometer, finally presses them together, and diminishes the capacity of the reservoir. It is partly for the sake of avoiding this inconvenience that we consider it good not to make an entire vacuum above the mercury, but to leave a portion of air in the tube, and at the same time to form a little reservoir at the summit of the instrument.

Differential Thermometer. — To graduate this instrument, you first maintain the two bulbs at an equal temperature, by which you determine the first fixed point, which is zero. Then, enveloping one of the two bulbs with melting snow, and elevating the other by means of a vessel with warm water, to a known temperature — to 20° Centigrade, for example — you fix a certain space, which you afterwards divide into 20 equal parts or degrees. The scale is continued by carrying successively to each side the known value of a degree.

GRADUATION OF RUMFORD'S THERMOSCOPE.

This instrument is graduated by dividing the tube which separates the two bulbs into equal parts, the number of which is arbitrary, though, in general, the thermoscope tube is divided into nine or eleven parts. There is always an odd number of degrees, and you manage so that the odd degree is found in the middle of the tube. It carries the mark of zero at each end, and the figures 1, 2, 3, &c. proceed from each end of this middle degree, and form two corresponding scales.

GRADUATION OF MARIOTTE'S TUBE.

You divide the little branch which is sealed at the end into a certain number of parts of equal capacity, and the large branch into inches and parts of inches. It is necessary to take care that the zero of the two ascending scales correspond, and are situated above the inferior bend formed by the two branches of the instrument.

THE END.

Lector House believes that a society develops through a two-fold approach of continuous learning and adaptation, which is derived from the study of classic literary works spread across the historic timeline of literature records. Therefore, we aim at reviving, repairing and redeveloping all those inaccessible or damaged but historically as well as culturally important literature across subjects so that the future generations may have an opportunity to study and learn from past works to embark upon a journey of creating a better future.

This book is a result of an effort made by Lector House towards making a contribution to the preservation and repair of original ancient works which might hold historical significance to the approach of continuous learning across subjects.

HAPPY READING & LEARNING!

LECTOR HOUSE LLP
E-MAIL: lectorpublishing@gmail.com